T0269598

The Endocannabinoidome
The World of Endocannabinoids and Related Mediators

The Endocannabinoidome
The World of Endocannabinoids
and Related Mediators

Vincenzo Di Marzo
Jenny Wang

AMSTERDAM • BOSTON • HEIDELBERG • LONDON
NEW YORK • OXFORD • PARIS • SAN DIEGO
SAN FRANCISCO • SINGAPORE • SYDNEY • TOKYO

Academic Press is an imprint of Elsevier

Academic Press is an imprint of Elsevier
32 Jamestown Road, London NW1 7BY, UK
525 B Street, Suite 1800, San Diego, CA 92101-4495, USA
225 Wyman Street, Waltham, MA 02451, USA
The Boulevard, Langford Lane, Kidlington, Oxford OX5 1GB, UK

Copyright © 2015 Elsevier Inc. All rights reserved.

No part of this publication may be reproduced or transmitted in any form or by any means,
electronic or mechanical, including photocopying, recording, or any information storage and
retrieval system, without permission in writing from the publisher. Details on how to seek
permission, further information about the Publisher's permissions policies and our arrangements
with organizations such as the Copyright Clearance Center and the Copyright Licensing
Agency, can be found at our website: www.elsevier.com/permissions.

This book and the individual contributions contained in it are protected under copyright by the
Publisher (other than as may be noted herein).

Notice
Knowledge and best practice in this field are constantly changing. As new research and experience
broaden our understanding, changes in research methods, professional practices, or medical
treatment may become necessary.

Practitioners and researchers must always rely on their own experience and knowledge in
evaluating and using any information, methods, compounds, or experiments described herein.
In using such information or methods they should be mindful of their own safety and the safety
of others, including parties for whom they have a professional responsibility.

To the fullest extent of the law, neither the Publisher nor the authors, contributors, or editors,
assume any liability for any injury and/or damage to persons or property as a matter of products
liability, negligence or otherwise, or from any use or operation of any methods, products,
instructions, or ideas contained in the material herein.

British Library Cataloguing-in-Publication Data
A catalogue record for this book is available from the British Library.

Library of Congress Cataloging-in-Publication Data
A catalog record for this book is available from the Library of Congress.

ISBN: 978-0-12-420126-2

For information on all Academic Press publications
visit our website at http://store.elsevier.com/

This book has been manufactured using Print On Demand technology. Each copy is produced to
order and is limited to black ink. The online version of this book will show color figures where
appropriate.

Working together
to grow libraries in
developing countries

www.elsevier.com • www.bookaid.org

DEDICATIONS

To my wife, Adriana, and my daughter, Marta, for bearing with my "obsession" with endocannabinoids and related mediators.

To the loving memory of J. Michael Walker, who, back in 2001, first pointed out to me the possibility of the existence of so many endocannabinoid-like mediators.

–Vincenzo Di Marzo

To my son Jeffrey, my parents, and my sister Xinxin and her husband Dan.

–Jenny Wang

TABLE OF CONTENTS

ACKNOWLEDGMENTS

To my coeditor, Dr. Jenny Wang, for having worked so hard and having played such an important role in the preparation of this book.

To the contributors of this book, who believe in the concept of the "endocannabinoidome" and have carried out fundamental research in this field.

To Daniele Piomelli and Vladimir Bezuglov, with whose important help I started working on bioactive fatty acid amides in the 1990s.

–Vincenzo Di Marzo

To David Woodward for his invaluable mentoring during the past decade.

To Vincenzo Di Marzo for getting me involved in editing this book with him. It started with an invitation for me to compile and edit an e-book on dermatology. So I checked with Vincenzo to seek his opinion. He suggested I coedit this book with him instead.

–Jenny Wang

Vincenzo Di Marzo, Jenny W. Wang

The discovery of two G-protein coupled receptors (GPCRs) for *Cannabis sativa* psychotropic principle Δ^9-tetrahydrocannabinol (THC), named "cannabinoid" receptors, and of their endogenous lipid ligands, the endocannabinoids [1], marked a milestone for both a century long series of mechanistic studies on the recreational/medicinal properties of this plant and its millennial history. At the turn of the century, these discoveries led to new successes as well as completely unpredicted findings, which can be summarized as follows: (1) the endogenous system composed of the cannabinoid CB1 and CB2 receptors, the endocannabinoids and the biochemical machinery to produce these lipids, also known as the "endocannabinoid system" (ECS), is one of the most pleiotropic signaling systems in vertebrates, by being involved in all aspects of mammalian physiology and pathology, and, for this same reason, it represents an attractive as well as very challenging target for the design and development of new therapeutic drugs [2]; (2) indeed, endocannabinoid-based drugs, such as rimonabant, have come to, and then gone from the market (whereas others, such as Sativex, are still successful and being actively proposed for more than one disease target [3,4]), albeit before the many nuances and complications of the ECS were fully understood [5]; and (3) the ECS is a complicated system, also because the endocannabinoids, like many lipid mediators: (i) are biosynthesized and degraded through redundant routes and enzymes that also participate in the regulation of the levels of other endogenous signals [6] and (ii) influence the activity also of noncannabinoid receptors [7].

The present book probably represents the first attempt to render a comprehensive overview of the complexity of the biochemistry and pharmacology of endocannabinoids and endocannabinoid-like mediators, which are suggested by some authors [8–10] as a new "ome" in its own right, i.e., the "endocannabinoidome." Indeed, from the first chapter, authored by Harald S. Hansen, Karen Kleberg, and Helle Adser Hassing, readers are reminded that, although the ability of the two major endocannabinoids, anandamide and 2-arachidonoylglycerol (2-AG), to activate CB1 and CB2 receptors was only discovered in the 1990s, these two compounds belong to two classes of metabolites, the N-acylethanolamines (NAEs) and the monoacylglycerols (MAGs), which had been previously known for decades as minor lipid constituents and metabolic intermediates with unknown function of both animal and plant organisms. Yet, the authors also discuss how these endocannabinoid "congeners", initially considered, at best, as mere "entourage" compounds, i.e., "accompanying" metabolites modulating the levels and/or action of their "more important" endocannabinoid brothers [11], have now been recognized as mediators in their own right with more or less specific targets among orphan GPCRs, nuclear receptors, and ion channels. Furthermore, as pointed out in the subsequent chapter by Jocelijn Meijerink, Michiel Balvers, Pierluigi Plastina, and Renger Witkamp, new members of the NAE family are starting to be revealed and investigated as it emerges that diets rich in n–3 polyunsaturated fatty acids, such as eicosapentaenoic and docosahexaenoic acid (particularly abundant in fish and krill oils), may lead to the accumulation of these congeners. As suggested by these, as well as other authors [12], some "omega-3" fatty acid amides might produce important anti-inflammatory and anticancer actions, again via as yet unidentified noncannabinoid receptor-mediated mechanisms.

The next step forward in the discovery of endocannabinoid-like mediators is discussed in the third chapter of the book by Emma Leishman and Heather B. Bradshaw, who describe how other N-acyl amides are found in tissues where they do not necessarily play a role as endocannabinoids (i.e., endogenous agonists of CB1 and CB2 receptors [13]). First and foremost, the N-acyl amino acids (also known as lipoaminoacids), such as the N-acyl glycines and N-acyl serines [14–16], which target transient receptor potential (TRP, such as vanilloid types 1–4) and voltage-activated (such as T-type Ca^{2+}) ion channels, as well as orphan GPCRs (such as GPR18); second, the N-acyl dopamines and N-acyl serotonins, together with their own targets and anabolic and catabolic mechanisms, which are the specific subject of Chapter 5, by Luciano De Petrocellis and one of the two coeditors of the book, Vincenzo Di Marzo. The overall impression provided by these two chapters is that, when it comes to define the molecular targets and metabolic routes of endocannabinoid-like mediators, a very high degree of promiscuity and redundancy emerges, which might hinder the translation of these findings to the clinical development of new therapies.

Chapter 4 by Lawrence J. Marnett, Philip J. Kingsley, and Daniel J. Hermanson is dedicated to discuss how arachidonic acid-containing members of these lipid families, particularly the endocannabinoids, can act as biosynthetic precursors of mediators obtained from the catalytic action of cyclooxygenase-2 and various prostaglandin synthases. The ensuing metabolites, i.e., the prostaglandin ethanolamides (or "prostamides") and the prostaglandin glycerol esters, are emerging as important mediators with possible feedback actions on the biological effects of their respective endocannabinoid precursors, exerted via receptors distinct from both cannabinoid and prostanoid receptors, whose molecular nature is still not fully understood. The comprehensively updated pharmacology of prostamide $F_{2\alpha}$ is then discussed in Chapter 6, by David F. Woodward and the other coeditor of this book, Jenny W. Wang. Thus, the discovery of endocannabinoids has branched both "vertically" and "horizontally" into the finding of often metabolically related bioactive fatty acid amides and esters, of which, however, to date we only appreciate in part their importance in biological functions.

What has been understood quite well is the biosynthetic routes and enzymes for some of the prostamides and all of the NAEs and MAGs, including anandamide and 2-AG, and the catabolic pathways of these latter lipid mediators. These achievements are comprehensively discussed in Chapter 7 by Kikuko Watanabe and David F. Woodward, on prostamide $F_{2\alpha}$, and in Chapter 8 by Natsuo Ueda, Kazuhito Tsuboi, and Toru Uyama, on NAEs and MAGs. This knowledge has already provided, and will provide even more in the future, the bases for the development of pharmacological (i.e., enzyme inhibitors) and genetic (i.e., "knockout" or "knockin" mice) tools, which will be crucial for the full understanding of the physiological and pathological roles of these lipid mediators, as will be the development of sensitive and accurate analytical methods for their measurement in tissues and biological fluids, a subject that is discussed in Chapter 9 by Fabiana Piscitelli.

Finally, perhaps the book could not be considered complete without an ideal "return" to the *Cannabis* plant, the root of all our knowledge of the endocannabinoidome. Indeed, in the last chapter of the book, Stephen P.H. Alexander nicely undertook the task of showing the readers how some of the emerging targets of endocannabinoid-like mediators are also shared by plant cannabinoids, and not just THC. In fact, some "phytocannabinoids" are now also being considered as pharmacologically relevant modulators of the same orphan GPCRs, nuclear receptors, and ion channels targeted by the fatty acid amides and glycerol esters, which are the protagonists of this book, and in many instances, by endocannabinoids as well.

In summary, we believe that the book manages to depict the complexity of the endocannabinoidome and its high potential in terms of future discoveries crucial for our understanding of how the biological functions of living organisms are regulated by lipid mediators, and, hence, for the identification of the strategies through which this new knowledge can be translated into the development of new medicines.

REFERENCES

[1] Mechoulam R, Fride E, Di Marzo V. Endocannabinoids. Eur J Pharmacol 1998;359(1):1–18.

[2] Di Marzo V, Bifulco M, De Petrocellis L. The endocannabinoid system and its therapeutic exploitation. Nat Rev Drug Discov 2004;3(9):771–84.

[3] Scheen AJ, Paquot N. Use of cannabinoid CB1 receptor antagonists for the treatment of metabolic disorders. Best Pract Res Clin Endocrinol Metab 2009;23(1):103–16.

[4] Sastre-Garriga J, Vila C, Clissold S, Montalban X. THC and CBD oromucosal spray (Sativex®) in the management of spasticity associated with multiple sclerosis. Expert Rev Neurother 2011;11(5):627–33.

[5] Di Marzo V. Targeting the endocannabinoid system: to enhance or reduce? Nat Rev Drug Discov 2008;7(5):438–55.

[6] Rahman IA, Tsuboi K, Uyama T, Ueda N. New players in the fatty acyl ethanolamide metabolism. Pharmacol Res 2014;86C:1–10.

[7] De Petrocellis L, Di Marzo V. Non-CB1, non-CB2 receptors for endocannabinoids, plant cannabinoids, and synthetic cannabimimetics: focus on G-protein-coupled receptors and transient receptor potential channels. J Neuroimmune Pharmacol 2010;5(1):103–21.

[8] Piscitelli F, Carta G, Bisogno T, Murru E, Cordeddu L, Berge K, et al. Effect of dietary krill oil supplementation on the endocannabinoidome of metabolically relevant tissues from high-fat-fed mice. Nutr Metab (Lond) 2011;8(1):51.

[9] Williams J, Pandarinathan L, Wood J, Vouros P, Makriyannis A. Endocannabinoid metabolomics: a novel liquid chromatography-mass spectrometry reagent for fatty acid analysis. AAPS J 2006;8(4):E655–60.

[10] Witkamp R, Meijerink J. The endocannabinoid system: an emerging key player in inflammation. Curr Opin Clin Nutr Metab Care 2014;17(2):130–8.

[11] Ben-Shabat S, Fride E, Sheskin T, Tamiri T, Rhee MH, Vogel Z, et al. An entourage effect: inactive endogenous fatty acid glycerol esters enhance 2-arachidonoyl-glycerol cannabinoid activity. Eur J Pharmacol 1998;353(1):23–31.

[12] Brown I, Cascio MG, Rotondo D, Pertwee RG, Heys SD, Wahle KW. Cannabinoids and omega-3/6 endocannabinoids as cell death and anticancer modulators. Prog Lipid Res 2013;52(1):80–109.

[13] Di Marzo V, Fontana A. Anandamide, an endogenous cannabinomimetic eicosanoid: 'killing two birds with one stone'. Prostaglandins Leukot Essent Fatty Acids 1995;53(1):1–11.

[14] Huang SM, Bisogno T, Petros TJ, Chang SY, Zavitsanos PA, Zipkin RE, et al. Identification of a new class of molecules, the arachidonyl amino acids, and characterization of one member that inhibits pain. J Biol Chem 2001;276(46):42639–44.

[15] Milman G, Maor Y, Abu-Lafi S, Horowitz M, Gallily R, Batkai S, et al. N-Arachidonoyl L-serine, an endocannabinoid-like brain constituent with vasodilatory properties. Proc Natl Acad Sci USA 2006;103(7):2428–33.

[16] Hanuš L, Shohami E, Bab I, Mechoulam R. N-Acyl amino acids and their impact on biological processes. Biofactors 2014 Apr. 21. doi: 10.1002/biof.1166. [Epub ahead of print].

CONTRIBUTORS

Stephen P.H. Alexander

Life Sciences, University of Nottingham Medical School, Nottingham, England

Michiel Balvers

Division of Human Nutrition, Wageningen University, EV Wageningen, The Netherlands

Heather B. Bradshaw

Department of Psychological and Brain Sciences, Indiana University, Bloomington, Indiana, USA

Luciano De Petrocellis

Endocannabinoid Research Group, Institute of Biomolecular Chemistry, Consiglio Nazionale delle Ricerche, Pozzuoli, Naples, Italy

Vincenzo Di Marzo

Endocannabinoid Research Group, Institute of Biomolecular Chemistry, Consiglio Nazionale delle Ricerche, Pozzuoli, Naples, Italy

Harald S. Hansen

Department of Drug Design and Pharmacology, Faculty of Health and Medical Sciences, University of Copenhagen, Copenhagen, Denmark

Helle Adser Hassing

Department of Drug Design and Pharmacology, Faculty of Health and Medical Sciences, University of Copenhagen, Copenhagen, Denmark

Daniel J. Hermanson

A.B. Hancock Jr. Memorial Laboratory for Cancer Research, Departments of Biochemistry, Chemistry, and Pharmacology, Vanderbilt Institute of Chemical Biology, Center in Molecular Toxicology, and Vanderbilt-Ingram Cancer Center, Nashville, Tennessee

Philip J. Kingsley

A.B. Hancock Jr. Memorial Laboratory for Cancer Research, Departments of Biochemistry, Chemistry, and Pharmacology, Vanderbilt Institute of Chemical Biology, Center in Molecular Toxicology, and Vanderbilt-Ingram Cancer Center, Nashville, Tennessee

Karen Kleberg

Department of Drug Design and Pharmacology, Faculty of Health and Medical Sciences, University of Copenhagen, Copenhagen, Denmark

Emma Leishman

Department of Psychological and Brain Sciences, Indiana University, Bloomington, Indiana, USA

Lawrence J. Marnett

A.B. Hancock Jr. Memorial Laboratory for Cancer Research, Departments of Biochemistry, Chemistry, and Pharmacology, Vanderbilt Institute of Chemical Biology, Center in Molecular Toxicology, and Vanderbilt-Ingram Cancer Center, Nashville, Tennessee

Jocelijn Meijerink

Division of Human Nutrition, Wageningen University, EV Wageningen, The Netherlands

Fabiana Piscitelli

Endocannabinoid Research Group, Institute of Biomolecular Chemistry, Consiglio Nazionale delle Ricerche, Pozzuoli, Naples, Italy

Pierluigi Plastina

Division of Human Nutrition, Wageningen University, EV Wageningen, The Netherlands; Department of Pharmaceutical Sciences, University of Calabria, Calabria, Italy

Kazuhito Tsuboi

Department of Biochemistry, Kagawa University School of Medicine, Kagawa, Japan

Natsuo Ueda

Department of Biochemistry, Kagawa University School of Medicine, Kagawa, Japan

Toru Uyama

Department of Biochemistry, Kagawa University School of Medicine, Kagawa, Japan

Jenny W. Wang

Department of Biological Sciences, Allergan Inc., Irvine, California

Kikuko Watanabe

Department of Nutrition, Koshien University, Takarazuka, Hyogo, Japan

Renger Witkamp

Division of Human Nutrition, Wageningen University, EV Wageningen, The Netherlands

David F. Woodward

Department of Biological Sciences, Allergan Inc., Irvine, California

CHAPTER *1*

Non-endocannabinoid *N*-Acylethanolamines and Monoacylglycerols: Old Molecules New Targets

Harald S. Hansen, Karen Kleberg, Helle Adser Hassing

1.1 INTRODUCTION

The non-endocannabinoid *N*-acylethanolamines (NAEs) and mono-acylglycerols (2-MAGs) owe their "non-endocannabinoid" name to their structural resemblance with the true endocannabinoids, anandamide (= *N*-arachidonoylethanolamine) and 2-arachidonoyl glycerol (2-AG) combined with their lack of effects on the two cannabinoid receptors. [1,2]. Anandamide and 2-AG have many endogenous structural analogs where arachidonic acid ($20:4n–6$) is substituted with shorter-chain unsaturated fatty acids such as linoleic acid ($18:2n–6$), oleic acid ($18:1n–9$), stearic acid ($18:0$), palmitic acid ($16:0$), or palmitoleic acid ($16:1n–7$), as the quantitatively most important. Contrary to the endocannabinoids, these NAEs and 2-MAGs cannot activate the cannabinoid receptors, but they do have biological activities in their own right and their synthesis and degradation

The Endocannabinoidome: The World of Endocannabinoids and Related Mediators. DOI: 10.1016/B978-0-12-420126-2.00001-8
Copyright © 2015 Elsevier Inc. All rights reserved

are in many cases coupled to those of the endocannabinoids. In fact, endogenous anandamide formation seems always to be associated with the formation of all the other NAEs such as oleoylethanolamide (OEA), palmitoylethanolamide (PEA), and linoleoylethanolamide (LEA) [3,4]. This co-formation seems not to be the case for 2-AG [5], which is probably formed by different enzymatic pathways than those forming other 2-MAGs such as 2-oleoyl glycerol (2-OG), 2-palmitoyl glycerol, and 2-linoleoylglycerol. The degradation of 2-AG and the other 2-MAGs is, however, mediated via the same enzymes.

Both groups of lipids are apparently found in all eukaryotic organisms, but while 2-MAG is abundantly and ubiquitously present as an intermediate in triacylglycerol (TAG) and phospholipid turnover [6–9], NAEs are less well-known, although they are also present in very small amounts in most eukaryotes such as yeast [10], slime molds [11], plants [12], insects [13], and mammals [14]. Although both the NAEs and the 2-MAGs have been known for more than 50 years as lipids in biological systems [15,16], their biology is still not fully understood and new physiological as well as pharmacological roles are currently revealing themselves at a high frequency. This chapter will cover the formation and possible biological roles of these non-endocannabinoid lipids in the mammalian physiology, as well as hinting at where new possible drug targets could be.

1.2 FORMATION AND DEGRADATION OF *N*-ACYLETHANOLAMINES

NAEs are formed within the tissues from their precursor phospholipids, the *N*-acyl-phosphatidylethanolamines and *N*-acyl-plasmalogens commonly abbreviated NAPEs (Figure 1.1). These NAPEs are found in small amounts in the membranes of eukaryotic cells (usually less than 0.05% of total phospholipids [17]). NAPEs are generated from phosphatidylethanolamine/plasmalogen and an acyl-donor phospholipid (often phosphatidylcholine) by a calcium-stimulated *N*-acyltransferase enzyme, which has not yet been purified or cloned [6,17], and by a group of poorly characterized enzymes from the phospholipase A/acyltransferase (PLA/AT) family [18,19]. It is not known whether formation of NAPEs occurs in any particular membrane structure within the cells, and it is unclear which enzymes are responsible for formation of NAPE

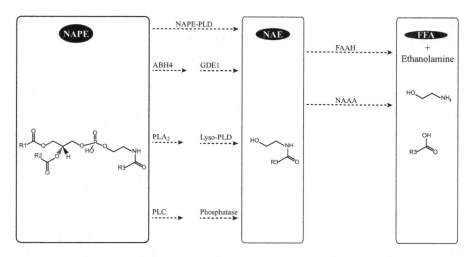

Fig. 1.1. Formation and degradation of NAE. N-Acyl-phosphatidylethanolamine (NAPE) can be converted to NAE by several enzymatic pathways. These pathways involve NAPE-hydrolyzing phospholipase D (NAPE-PLD), α/β-hydrolase-4 (ABH4) – glycerophosphodiester phosphodiesterase-1 (GDE1), phospholipase A₂ (PLA₂) – lysophospholipase D (lyso-PLD), and phospholipase C (PLC) – phosphatase. Degradation of NAE to ethanolamine and free fatty acid (FFA) can be catalyzed by FAAH and NAAA. R1–R3: long chain alkyl group.

in different tissues. Apparently, there is no acyl group selectivity in the *N*-acyltransferase reaction, entailing that it is the fatty acid in the *sn*-1 position of the acyl-donor phospholipid, which ends up in the *N*-acyl position of NAPE [6,17]. There are several enzymes involved in the formation of NAEs (Figure 1.1), and the best known is NAPE-hydrolyzing phospholipase D (NAPE-PLD), which has been cloned and well characterized [20]. Tissue levels of non-endocannabinoid NAEs (stearoylethanolamide (SEA), OEA, PEA, and LEA) are generally in the order of picomole per-gram tissue, and usually in the following quantitative order: SEA = PEA > OEA > LEA, except in the small intestine, where LEA may be present at the highest level [21–23].

Mice lacking NAPE-PLD have decreased tissue levels of NAEs, although it is unclear whether the decrease is restricted only to saturated and monounsaturated *N*-acylethanolamines [24] or also includes anandamide and polyunsaturated NAEs [25]. The fact that *N*-acylethanolmines are not entirely absent in the knockout mice indicates that other biosynthetic pathways exist as indicated in Figure 1.1, and as discussed in Chapter 8. NAEs are hydrolyzed within the tissues by mainly two enzymes, fatty acid amide hydrolase (FAAH) [26] and *N*-acylethanolamine acid amidase (NAAA)

[27]. FAAH, which is located in the endoplasmic reticulum, exhibits the highest enzymatic activity for anandamide and other unsaturated NAEs; while NAAA, which mainly is located in lysosomes, primarily hydrolyzes saturated NAEs such as PEA. Mice lacking FAAH have increased tissue levels of all NAEs including anandamide [28], stressing the role of this enzyme in the regulation of NAE levels. The consequences of NAAA elimination in a knockout mouse model is yet to be elucidated.

1.3 BIOLOGICAL ROLES AND DRUG TARGETS OF NON-ENDOCANNABINOID N-ACYLETHANOLAMINES

OEA, PEA, and LEA can with different potencies activate the transcription factor PPARα [23,29,30], the G-protein-coupled receptor GPR119 [31,32], the vanilloid receptor [33,34], and several ion channels, which may also be activated by true endocannabinoids [35–37]. Intraperitoneal injection or oral administration of these NAEs into experimental animals have revealed different effects, including inhibition of food intake (OEA, PEA, LEA) [23,38,39], inhibition of inflammation (PEA) [40–45], inhibition of pain (OEA, PEA) [46–48], inhibition of atherosclerosis (OEA) [49], as well as anticonvulsive (PEA) [50], and neuroprotective effects (PEA, OEA) [51–55]. A number of clinical studies have shown pain-relieving effect of PEA, concomitant reduction in disability, and/or improvement of neurological functions and quality of life using a dietary supplement of PEA with the trade names Normast®, Pelvilen®, or Glialia® [56]. Most of the studies were small non-controlled studies but there were also a few larger placebo-controlled studies [56]. Furthermore, a randomized, double-blind placebo-controlled crossover study with 40 ocular hypertensive patients and 40 controls showed that oral PEA induced a reduction in intraocular pressure and improvement of flow-mediated vasodilation [57]. It seems clear that there is a potential of using these non-endocannabinoid NAEs or synthetic derivatives as drug candidates in a number of human pathologies.

Whether endogenous OEA, PEA, and LEA also have all of the above-mentioned effects is less certain, although it is known that these signaling lipids will increase in cases of tissue injury and inflammation, potentially as a protective mechanism [5,58].

The use of FAAH inhibitors, which will increase the tissue levels of OEA, PEA, and LEA, will also increase the tissue levels of anandamide.

FAAH inhibitors have shown many positive biological effects in animal models of human pathologies, but most of these effects may primarily be due to an increase in the endogenous levels of anandamide followed by activation of cannabinoid receptors [59–63]. However, transiently up-regulation of NAPE-PLD and thereby of the endogenous level of OEA in the small intestine by the use of an adenoviral vector [64] suggests that endogenous OEA may have a role in the regulation of food intake [65]. Prolonged intake of a diet high in fat is known to decrease intestinal levels of OEA, PEA, and LEA [23,66,67], and this may contribute to the well-known hyperphagic effect of sustained high dietary fat-levels [68].

The hypothesis of NAE's importance in dietary fat sensing is supported by recent evidence, which strengthen the role of OEA as a pivotal intestinal signaling molecule in central appetite regulation. The decreased intestinal OEA levels observed in diet-induced obese mice were correlated to an abolished increase in central dopamine levels, which is otherwise prompted by intragastric fat infusions in mice [67]. Importantly, exogenous OEA was shown to reestablish the increase in central dopamine levels in the obese mice, and effects on both appetite and food preference were observed, pointing to OEA as an important homeostatic link between intestinal fat processing and central appetite regulation. The increased dopamine levels in response to fat infusions were dependent on PPARα and the vagus nerve [67], which point to PPARα as a potential target in obesity management. PPARα, however, is a transcription factor with many implications, which differ among species such as rodents, lagomorphs, and primates, and clinical use of PPARα agonists such as fenofibrate is not associated with a decline in body weight [69,70].

NAAA inhibitors will increase the endogenous levels of PEA and other nonpolyunsaturated NAEs [71]. These inhibitors have anti-inflammatory and anti-nociceptive effects in rodent disease models [71–73]. Thus, NAAA is also a drug target.

1.4 FORMATION AND DEGRADATION OF NON-ENDOCANNABINOID 2-MONOACYLGLYCEROLS

2-MAGs are intermediates in TAG and phospholipid metabolism [6]. Under normal physiological conditions, 2-MAGs are most commonly formed during digestion of dietary fat [68,74], hydrolysis of lipoprotein

TAG by lipoprotein lipase (LPL) [6], and lipolysis of stored TAG in the tissues [6]. These pathways are distinct from the major biosynthetic pathway for 2-AG formation, which is via diacylglycerols derived from the turnover of phosphoinositides [5]. The inositol phospholipids generally contain stearic acid in *sn*-1 position and arachidonic acid in *sn*-2 position, which is why the other noncannabinoid 2-MAGs usually do not originate from this pathway. However, in some rapid developing cancers, oleic acid may be found in the *sn*-2 position of phosphatidylinositol and thereby potentially give rise to formation of 2-OG [75] (Figure 1.2).

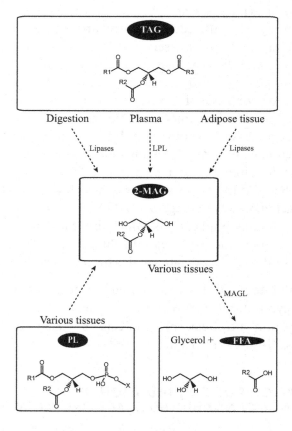

Fig. 1.2. Formation and degradation of 2-MAG. 2-MAG is formed in the intestine from ingested TAG via action of different digestive lipases, in the plasma by LPL, or in adipose tissue during lipolysis. In various tissues, 2-MAG may be formed from phospholipids (PL). The main route of degradation is via action of MAGL. R1–R3: long chain alkyl group. X: glycerol, ethanol, choline, serine, or inositol.

1.5 BIOLOGICAL ROLES AND DRUG TARGETS OF NON-ENDOCANNABINOID 2-MONOACYLGLYCEROLS

The role of 2-MAG as a signaling molecule is relatively new [32], and the molecule has often been neglected or completely overshadowed by the fatty acids generated from lipolysis. In general, it has been viewed as an inert intermediate in synthesis and degradation of acylglycerols. Lately, new evidence reveals possible signaling properties of 2-MAGs and indicates that it may be involved in intestinal lipid sensing via activation of GPR119 and release of the incretin hormone GLP-1 [32,74] (Figure 1.3), and may also mediate insulin secretion in the pancreas. In the pancreas, it may originate from intracellular TAG turnover and it may function through activation of GPR119 [74], or through directly stimulation of the exocytosis mechanism [76].

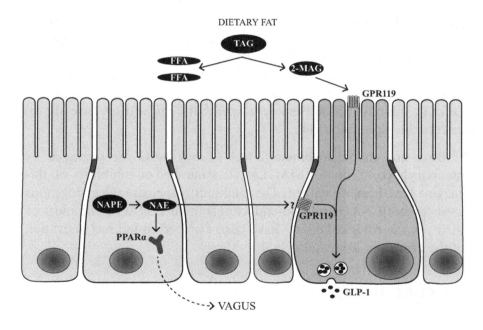

Fig. 1.3. Proposed actions of 2-MAG and NAEs in the intestinal mucosa. Dietary fat (TAG) is hydrolyzed in the intestinal lumen to FFAs and 2-MAG. 2-MAG is then available for activation of GPR119 receptors on the apical membrane of enteroendocrine cells and mediate basolateral release of GLP-1. NAEs are synthesized inside the enterocyte from NAPE and can activate PPARα within the same cells or possibly activate GPR119 on neighboring enteroendocrine cells via a paracrine route. The anorectic effect of NAE-activated PPARα is mediated via the vagus nerve.

Some of the positive metabolic effects of the gut microbiota may be explained by their effect on enhancing intestinal levels of 2-OG and stimulating GLP-1 release [77]. GPR119 is already a drug target for potential treatment of diabetes type 2 [74,78]. Since GPR119 detects the dietary fat-derived 2-MAG in the intestinal lumen, it has been suggested to be a fat sensor. Dietary fat is a potent stimulant of intestinal GLP-1 release in humans [79], but an increase in dietary fat in order to treat diabetes type 2 is not realistic. However, structured TAG with short-chain fatty acids in the sn-1 and sn-3 position could serve as a sort of fat substitute, i.e., fat having lower energy content, and serving as a pro-drug for 2-OG release in the intestinal lumen. Such structured TAGs are able to give a GLP-1 release in humans comparable to an equimolar amount of olive oil [80].

In the enterocyte during fat digestion, 2-MAG is mostly converted to diacylglycerol and TAG. Inhibition of this acylation, whether it is of monoacylglycerol acyltransferase (MGAT) or of diacylglycerol acyltransferase (DGAT), may lead to accumulation of 2-OG and potentially improved GLP-1 secretion. Mice deficient in the intestine-specific MGAT2 and DGAT1 have higher postprandial plasma levels of GLP-1 [81,82], suggesting that inhibitors of these enzymes could be drug targets for the purpose of improving metabolic diseases [83]. A number of DGAT1 inhibitors have been developed but clinical trials have shown undesirable side effects [84].

Both non-endocannabinoid 2-MAGs and 2-AG are degraded by monoacylglycerol lipase (MAGL) and a number of inhibitors of this enzyme have been developed. These inhibitors increase the endogenous levels of both 2-AG and non-endocannabinoid 2-MAGs, and most of their pharmacological effects have been ascribed to indirect activation of the cannabinoid receptors [85].

1.6 CONCLUSION

The non-endocannabinoid NAEs and 2-MAGs hold important biological functions ranging from anti-inflammatory, antinociceptive, and anorectic effects to promoting GLP-1 and possibly also insulin secretion. In addition to the direct drug potentials of these molecules or their synthetic analogs, the enzymes involved in their degradation are also drug targets.

ACKNOWLEDGMENT

Research in the authors' laboratory has been supported by grants from the UNIK programme Food, Fitness & Pharma, from the Augustinus Foundation, and the Novonordisk Foundation.

REFERENCES

[1] Maccarrone M, Gasperi V, Catani MV, Diep TA, Dainese E, Hansen HS, et al. The endocannabinoid system and its relevance for nutrition. Annu Rev Nutr 2010;30:423–40.

[2] Pertwee RG, Howlett AC, Abood ME, Alexander SP, Di Marzo V, Elphick MR, et al. International Union of Basic and Clinical Pharmacology. LXXIX. Cannabinoid Receptors and Their Ligands: Beyond CB1 and CB2. Pharmacol Rev 2010;62:588–631.

[3] Hansen HS. Palmitoylethanolamide and other anandamide congeners. Proposed role in the diseased brain. Exp Neurol 2010;224:48–55.

[4] Ueda N, Tsuboi K, Uyama T. Enzymological studies on the biosynthesis of *N*-acylethanolamines. Biochim Biophys Acta 2010;1801:1274–85.

[5] Ueda N, Tsuboi K, Uyama T, Ohnishi T. Biosynthesis and degradation of the endocannabinoid 2-arachidonoylglycerol. BioFactors 2010;37:1–7.

[6] Young SG, Zechner R. Biochemistry and pathophysiology of intravascular and intracellular lipolysis. Genes Dev 2013;27:459–84.

[7] Richmond GS, Smith TK. Phospholipases A1. Int J Mol Sci 2011;12:588–612.

[8] Hung ND, Kim MR, Sok DE. Purification and characterization of lysophospholipase C from pig brain. Neurochem Res 2010;35:50–9.

[9] Li J, Dong Y, Lu X, Wang L, Peng W, Zhang XC, et al. Crystal structures and biochemical studies of human lysophosphatidic acid phosphatase type 6. Protein Cell 2013;4:548–61.

[10] Muccioli GG, Sia A, Muchowski PJ, Stella N. Genetic manipulation of palmitoyl-ethanolamide production and inactivation in *Saccharomyces cerevisiae*. PLoS ONE 2009;4:e5942.

[11] Hayes AC, Stupak J, Li J, Cox AD. Identification of *N*-acylethanolamines in *Dictyostelium discoideum* and confirmation of their hydrolysis by fatty acid amide hydrolase. J Lipid Res 2013;54:457–66.

[12] Blancaflor EB, Kilaru A, Keereetaweep J, Khan BR, Faure L, Chapman KD. *N*-Acyl-ethanolamines: lipid metabolites with functions in plant growth and development. Plant J 2014, in press.

[13] Fezza F, Dillwith JW, Bisogno T, Tucker JS, Di Marzo V, Sauer JR. Endocannabinoids and related fatty acid amides, and their regulation, in the salivary glands of the lone star tick. Biochim Biophys Acta Mol Cell Biol Lipids 2003;1633:61–7.

[14] Hansen HS, Diep TA. *N*-Acylethanolamines, anandamide and food intake. Biochem Pharmacol 2009;78:553–60.

[15] Mattson FH, Benedict JH, Martin JB, Beck LW. Intermediates formed during the digestion of triglycerides. J Nutr 1952;48:335–44.

[16] Kuehl Jr FA, Jacob TA, Ganley OH, Ormond RE, Meisinger MAP. The identification of *N*-(2-hydroxyethyl)-palmitamide as a naturally occurring anti-inflammatory agent. J Am Oil Chem Soc 1957;79:5577–8.

[17] Wellner N, Diep TA, Janfelt C, Hansen HS. *N*-Acylation of phosphatidyl-ethanolamine and its biological functions in mammals. Biochim Biophys Acta 2013;1831:652–62.

[18] Uyama T, Inoue M, Okamoto Y, Shinohara N, Tai T, Tsuboi K, et al. Involvement of phospholipase A/acyltransferase-1 in N-acylphosphatidylethanolamine generation. Biochim Biophys Acta 2013;1831:1690–701.

[19] Uyama T, Ikematsu N, Inoue M, Shinohara N, Jin XH, Tsuboi K, et al. Generation of N-acylphosphatidylethanolamine by members of the phospholipase A/acyltransferase (PLA/AT) family. J Biol Chem 2012;287:31905–19.

[20] Okamoto Y, Morishita J, Tsuboi K, Tonai T, Ueda N. Molecular characterization of a phospholipase D generating anandamide and its congeners. J Biol Chem 2004;279: 5298–305.

[21] Hansen HS, Moesgaard B, Hansen HH, Petersen G. N-Acylethanolamines and precursor phospholipids – relation to cell injury. Chem Phys Lipids 2000;108:135–50.

[22] Hansen HS. Effect of diet on tissue levels of palmitoylethanolamide. CNS Neurol Disord Drug Targets 2013;12:17–25.

[23] Diep TA, Madsen AN, Holst B, Kristiansen MM, Wellner N, Hansen SH, et al. Dietary fat decreases intestinal levels of the anorectic lipids through a fat sensor. FASEB J 2011;25: 765–74.

[24] Leung D, Saghatelian A, Simon GM, Cravatt BF. Inactivation of N-acyl-phosphatidyletha-nolamine phospholipase D reveals multiple mechanisms for the biosynthesis of endocannabi-noids. Biochemistry 2006;45:4720–6.

[25] Tsuboi K, Okamoto Y, Ikematsu N, Inoue M, Shimizu Y, Uyama T, et al. Enzymatic formation of N-acyl-ethanolamines from N-acylethanolamine plasmalogen through N-acylphospha-tidylethanolamine-hydrolyzing phospholipase D-dependent and -independent pathways. Biochim Biophys Acta 2011;1811:565–77.

[26] Long JZ, Cravatt BF. The metabolic serine hydrolases and their functions in Mammalian physiology and disease. Chem Rev 2011;111:6022–63.

[27] Ueda N, Tsuboi K, Uyama T. N-Acylethanolamine metabolism with special reference to N-acylethanolamine-hydrolyzing acid amidase (NAAA). Prog Lipid Res 2010;49:299–315.

[28] Cravatt BF, Demarest K, Patricelli MP, Bracey MH, Giang DK, Martin BR, et al. Supersen-sitivity to anandamide and enhanced endogenous cannabinoid signaling in mice lacking fatty acid amide hydrolase. Proc Natl Acad Sci USA 2001;98:9371–6.

[29] Fu J, Gaetani S, Oveisi F, LoVerme J, Serrano A, Rodríguez de Fonseca F, et al. Oleoylethanol-amide regulates feeding and body weight through activation of the nuclear receptor PPARα. Nature 2003;425:90–3.

[30] Lo Verme J, Fu J, Astarita G, La Rana G, Russo R, Calignano A, et al. The nuclear receptor peroxisome proliferator-activated receptor-α mediates the anti-inflammatory actions of pal-mitoylethanolamide. Mol Pharmacol 2005;67:15–9.

[31] Overton HA, Babbs AJ, Doel SM, Fyfe MC, Gardner LS, Griffin G, et al. Deorphanization of a G protein-coupled receptor for oleoylethanolamide and its use in the discovery of small-molecule hypophagic agents. Cell Metab 2006;3:167–75.

[32] Hansen KB, Rosenkilde MM, Knop FK, Wellner N, Diep TA, Rehfeld JF, et al. 2-Oleoyl glycerol is a GPR119 agonist and signals GLP-1 release in humans. J Clin Endocrinol Metab 2011;96:E1409–17.

[33] Smart D, Gunthorpe MJ, Jerman JC, Nasir S, Gray J, Muir AI, et al. The endogenous lip-id anandamide is a full agonist at the human vanilloid receptor (hVR1). Br J Pharmacol 2000;129:227–30.

[34] Movahed P, Jönsson BAG, Birnir B, Wingstrand JA, Jorgensen TD, Ermund A, et al. Endog-enous unsaturated C18 N-acylethanolamines are vanilloid receptor (TRPV1) agonists. J Biol Chem 2005;280:38496–504.

[35] Voitychuk OI, Asmolkova VS, Gula NM, Sotkis GV, Galadari S, Howarth FC, et al. Modula-
      tion of excitability, membrane currents and survival of cardiac myocytes by N-acylethanol-
      amines. Biochim Biophys Acta 2012;182:1167–76.

[36] Amoros I, Barana A, Caballero R, Gomez R, Osuna L, Lillo MP, et al. Endocannabinoids
      and cannabinoid analogues block human cardiac Kv4.3 channels in a receptor-independent
      manner. J Mol Cell Cardiol 2010;48:201–10.

[37] Barana A, Amoros I, Caballero R, Gomez R, Osuna L, Lillo MP, et al. Endocannabinoids and
      cannabinoid analogues block cardiac hKv1.5 channels in a cannabinoid receptor-independent
      manner. Cardiovasc Res 2010;85:56–67.

[38] Rodríguez de Fonseca F, Navarro M, Gómez R, Escuredo L, Nava F, Fu J, et al. An anorexic
      lipid mediator regulated by feeding. Nature 2001;414:209–12.

[39] Thabuis C, Destaillats F, Landrier JF, Tissot-Favre D, Martin JC. Analysis of gene expression
      pattern reveals potential targets of dietary oleoylethanolamide in reducing body fat gain in
      C3H mice. J Nutr Biochem 2010;21:922–8.

[40] Esposito G, Capoccia E, Turco F, Palumbo I, Lu J, Steardo A, et al. Palmitoylethanolamide
      improves colon inflammation through an enteric glia/toll like receptor 4-dependent PPAR-
      alpha activation. Gut 2014;68:1300–12.

[41] Impellizzeri D, Esposito E, Di Paola R, Ahmad A, Campolo M, Peli A, et al. Palmitoyletha-
      nolamide and luteolin ameliorate development of arthritis caused by injection of collagen type
      II in mice. Arthritis Res Ther 2013;15:R192.

[42] Cerrato S, Brazis P, Della Valle MF, Miolo A, Petrosin S, Di Marzo V, et al. Effects of palmi-
      toylethanolamide on the cutaneous allergic inflammatory response in Ascaris hypersensitive
      Beagle dogs. Vet J 2012;191:377–82.

[43] Petrosino S, Cristino L, Karsak M, Gaffal E, Ueda N, Tuting T, et al. Protective role of palmi-
      toylethanolamide in contact allergic dermatitis. Allergy 2011;65:698–711.

[44] Mattace RG, Simeoli R, Russo R, Santoro A, Pirozzi C, de Villa Bianca d'E, et al.
      N-Palmitoylethanolamide protects the kidney from hypertensive injury in spontaneously
      hypertensive rats via inhibition of oxidative stress. Pharmacol Res 2013;76C:67–76.

[45] Mazzari S, Canella R, Petrelli L, Marcolongo G, Leon A. N-(2-Hydroxyethyl)-hexadecana-
      mide is orally active in reducing edema formation and inflammatory hyperalgesia by down-
      modulating mast cell activation. Eur J Pharmacol 1996;300:227–36.

[46] Calignano A, La Rana G, Piomelli D. Antinociceptive activity of the endogenous fatty acid
      amide, palmitylethanolamide. Eur J Pharmacol 2001;419:191–8.

[47] Suardiaz M, Estivill-Torrus G, Goicoechea C, Bilbao A, Rodríguez de Fonseca F. Analgesic prop-
      erties of oleoylethanolamide (OEA) in visceral and inflammatory pain. Pain 2007;33:99–110.

[48] Di Cesare ML, D'Agostino G, Pacini A, Russo R, Zanardelli M, Ghelardini C, et al.
      Palmitoylethanolamide is a disease-modifying agent in peripheral neuropathy: pain relief
      and neuroprotection share a PPAR-alpha-mediated mechanism. Mediators 2013;2013:328797.

[49] Fan A, Wu X, Wu H, Li L, Huang R, Zhu Y, et al. Atheroprotective effect of oleoylethanol-
      amide (OEA) targeting oxidized LDL. PLoS ONE 2014;9:e85337.

[50] Lambert DM, Vandevoorde S, Diependaele G, Govaerts SJ, Robert AR. Anti-convulsant activ-
      ity of N-palmitoylethanolamide, a putative endocannabinoid, in mice. Epilepsia 2001;42:321–7.

[51] Ahmad A, Genovese T, Impellizzeri D, Crupi R, Velardi E, Marino A, et al. Reduction of isch-
      emic brain injury by administration of palmitoylethanolamide after transient middle cerebral
      artery occlusion in rats. Brain Res 2012;1477:45–58.

[52] D'Agostino G, Russo R, Avagliano C, Cristiano C, Meli R, Calignano A. Palmitoyl-ethanol-
      amide protects against the amyloid-beta25-35-induced learning and memory impairment in mice,
      an experimental model of Alzheimer disease. Neuropsychopharmacology 2012;37:1784–92.

[53] Zhou Y, Yang L, Ma A, Zhang X, Li W, Yang W, et al. Orally administered oleoylethanol-amide protects mice from focal cerebral ischemic injury by activating peroxisome proliferator-activated receptor alpha. Neuropharmacology 2012;63:242–9.

[54] Galan-Rodriguez B, Suarez J, Gonzalez-Aparicio R, Bermudez-Silva FJ, Maldonado R, Robledo P, et al. Oleoylethanolamide exerts partial and dose-dependent neuroprotection of substantia nigra dopamine neurons. Neuropharmacology 2009;56:653–64.

[55] Garg P, Duncan RS, Kaja S, Zabaneh A, Chapman KD, Koulen P. Lauroyl-ethanolamide and linoleoylethanolamide improve functional outcome in a rodent model for stroke. Neurosci Lett 2011;492:134–8.

[56] Skaper SD, Facci L, Fusco M, Della Valle MF, Zusso M, Costa B, et al. Palmitoylethanol-amide, a naturally occurring disease-modifying agent in neuropathic pain. Inflammopharma-cology 2014;22:79–94.

[57] Strobbe E, Cellini M, Campos EC. Effectiveness of palmitoylethanolamide on endothelial dysfunction in ocular hypertensive patients: a randomized, placebo-controlled cross-over study. Invest Ophthalmol Vis Sci 2013;54:968–73.

[58] Buczynski MW, Svensson CI, Dumlao DS, Fitzsimmons BL, Shim JH, Scherbart TJ, et al. Inflammatory hyperalgesia induces essential bioactive lipid production in the spinal cord. J Neurochem 2010;114:981–93.

[59] Booker L, Kinsey SG, Abdullah RA, Blankman JL, Long JZ, Ezzili C, et al. The FAAH inhibitor PF-3845 acts in the nervous system to reverse lipopolysaccharide-induced tactile al-lodynia in mice. Br J Pharmacol 2012;165:2485–96.

[60] Kawahara H, Drew G, Christie M, Vaughan C. Inhibition of fatty acid amide hydrolase un-masks CB(1) receptor and TRPV1 channel-mediated modulation of glutamatergic synaptic transmission in midbrain periaqueductal grey. Br J Pharmacol 2011;163:1214–22.

[61] Mukhopadhyay P, Horvath B, Rajesh M, Matsumoto S, Saito K, Batkai S, et al. Fatty acid amide hydrolase is a key regulator of endocannabinoid-induced myocardial tissue injury. Free Radic Biol Med 2011;50:179–95.

[62] Clapper JR, Moreno-Sanz G, Russo R, Guijarro A, Vacondio F, Duranti A, et al. Anan-damide suppresses pain initiation through a peripheral endocannabinoid mechanism. Nat Neurosci 2010;13:1265–70.

[63] Ahn K, Johnson DS, Mileni M, Beidler D, Long JZ, McKinney MK, et al. Discovery and characterization of a highly selective FAAH inhibitor that reduces inflammatory pain. Chem Biol 2009;16:411–20.

[64] Fu J, Kim J, Oveisi F, Astarita G, Piomelli D. Targeted enhancement of oleoylethanolamide production in proximal small intestine induces across-meal satiety in rats. Am J Physiol Regul Integr Comp Physiol 2008;295:R45–50.

[65] Piomelli D. A fatty gut feeling. Trends Endocrinol Metab 2013;24:332–41.

[66] Artmann A, Petersen G, Hellgren LI, Boberg J, Skonberg C, Hansen SH, et al. Influence of dietary fatty acids on endocannbinoid and n-acylethanolamine levels in rat brain, liver and small intestine. Biochim Biophys Acta Mol Cell Biol Lipids 2008;1781:200–12.

[67] Tellez LA, Medina S, Han W, Ferreira JG, Licona-Limon P, Ren X, et al. A gut lipid messen-ger links excess dietary fat to dopamine deficiency. Science 2013;341:800–2.

[68] Kleberg K, Hassing HA, Hansen HS. Classical endocannabinoid-like compounds and their regulation by nutrients. BioFactors 2014, in press.

[69] Hennuyer N, Poulain P, Madsen L, Berge RK, Houdebine LM, Branellec D, et al. Beneficial effects of fibrates on apolipoprotein A-I metabolism occur independently of any peroxisome proliferative response. Circulation 1999;99:2445–51.

[70] Winegar DA, Brown PJ, Wilkison WO, Lewis MC, Ott RJ, Tong WQ, et al. Effects of fenofi-brate on lipid parameters in obese rhesus monkeys. J Lipid Res 2001;42:1543–51.

[71] Sasso O, Moreno-Sanz G, Martucci C, Realini N, Dionisi M, Mengatto L, et al. Antinociceptive effects of the *N*-acylethanolamine acid amidase inhibitor ARN077 in rodent pain models. Pain 2013;154:350–60.

[72] Ponzano S, Bertozzi F, Mengatto L, Dionisi M, Armirotti A, Romeo E, et al. Synthesis and structure-activity relationship (SAR) of 2-methyl-4-oxo-3-oxetanylcarbamic acid esters, a class of potent *N*-acylethanolamine acid amidase (NAAA) inhibitors. J Med Chem 2013;56:6917–34.

[73] Solorzano C, Zhu C, Battista N, Astarita G, Lodola A, Rivara S, et al. Selective *N*-acylethanolamine-hydrolyzing acid amidase inhibition reveals a key role for endogenous palmitoylethanolamide in inflammation. Proc Natl Acad Sci USA 2009;106:20966–71.

[74] Hansen HS, Rosenkilde MM, Holst JJ, Schwartz TW. GPR119 as a fat sensor. Trends Pharmacol Sci 2012;33:374–81.

[75] Kawashima M, Iwamoto N, Kawaguchi-Sakita N, Sugimoto M, Ueno T, Mikami Y, et al. High-resolution imaging mass spectrometry reveals detailed spatial distribution of phosphatidylinositols in human breast cancer. Cancer Sci 2013;104:1372–9.

[76] Prentki M, Matschinsky FM, Madiraju SR. Metabolic signaling in fuel-induced insulin secretion. Cell Metab 2013;18:162–85.

[77] Everard A, Belzer C, Geurts L, Ouwerkerk JP, Druart C, Bindels LB, et al. Cross-talk between *Akkermansia muciniphila* and intestinal epithelium controls diet-induced obesity. Proc Natl Acad Sci USA 2012;110:9066–71.

[78] Kang SU. GPR119 agonists: a promising approach for T2DM treatment? A SWOT analysis of GPR119. Drug Discov Today 2013;18:1309–15.

[79] Carr RD, Larsen MO, Winzell MS, Jelic K, Lindgren O, Deacon CF, et al. Incretin and islet hormonal responses to fat and protein ingestion in healthy men. Am J Physiol Endocrinol Metab 2008;295:E779–84.

[80] Mandøe MJ, Hansen KB, Knop FK, Holst JJ, Hansen HS. Incretin hormones in patients with type 2 diabetes are increased by diet-oil, a pro-drug for GPR119 receptor agonist, 2-oleoyl-glycerol. Am Diab Assoc 2012;. abstract 2012.

[81] Okawa M, Fujii K, Ohbuchi K, Okumoto M, Aragane K, Sato H, et al. Role of MGAT2 and DGAT1 in the release of gut peptides after triglyceride ingestion. Biochem Biophys Res Commun 2009;390:377–81.

[82] Yen CL, Cheong ML, Grueter C, Zhou P, Moriwaki J, Wong JS, et al. Deficiency of the intestinal enzyme acyl CoA:monoacylglycerol acyltrans-ferase-2 protects mice from metabolic disorders induced by high-fat feeding. Nat Med 2009;15:442–6.

[83] Lin HV, Chen D, Shen Z, Zhu L, Ouyang X, Vongs A, et al. Diacylglycerol acyltransferase-1 (DGAT1) inhibition perturbs postprandial gut hormone release. PLoS ONE 2013;8:e54480.

[84] Devita RJ, Pinto S. Current status of the research and development of diacylglycerol O-acyltransferase 1 (DGAT1) inhibitors. J Med Chem 2013;56:9820–5.

[85] Pertwee RG. Elevating endocannabinoid levels: pharmacological strategies and potential therapeutic applications. Proc Nutr Soc 2014;73:96–105.

CHAPTER 2

Omega-3 Polyunsaturated *N*-Acylethanolamines: A Link Between Diet and Cellular Biology

Jocelijn Meijerink, Michiel Balvers, Pierluigi Plastina, Renger Witkamp

ABBREVIATIONS

AA arachidonic acid
AEA *N*-arachidonoylethanolamine (anandamide)
2-AG 2-arachidonoylglycerol
ALA α-linolenic acid (18:3*n*–3)

The Endocannabinoidome: The World of Endocannabinoids and Related Mediators. DOI: 10.1016/B978-0-12-420126-2.00002-X
Copyright © 2015 Elsevier Inc. All rights reserved

CB (receptor) cannabinoid (receptor)
COX cyclooxygenase
CREB cAMP response element binding protein
DHA docosahexaenoic acid (22:6n–3)
DHEA N-docosahexaenoylethanolamine
DPA docosapentaenoic acid (22;5n–3)
ECS endocannabinoid system
EPEA N-eicosapentaenoylethanolamine
FAAH fatty acid amide hydrolase
MAP2 microtubule-associated protein 2
NAEs N-acylethanolamines
NO nitric oxide
NSC neural stem cell
OEA N-oleoylethanolamine
PEA N-palmitoylethanolamine
PKA protein kinase A
PPAR peroxisome proliferator-activated receptor
(LC-)PUFA (long chain-)polyunsaturated fatty acid
TRPA1 transient receptor potential ankyrin 1
TRVP1 transient receptor potential channel type V1

2.1 INTRODUCTION

According to the IUPHAR classification system, endocannabinoids are defined as endogenous compounds capable of binding and functionally activating the cannabinoid receptors CB_1 and CB_2 [1,2]. Currently, at least nine of these "classical" endocannabinoids are known, which are all derived from n-6 long chain (C18 or longer) polyunsaturated fatty acids (LC-PUFAs), or from oleic acid (C18:1) [3]. This number might increase in the future, as for example, two n-3 LC-PUFA-derived ethanolamides, N-eicosapentaenoylethanolamine (EPEA) and N-docosahexaenoylethanolamine (DHEA) were also shown to bind to and activate CB receptors, be it with lower affinity than, for example, anandamide [4,5]. CB receptors, their endogenous ligands, and enzymes involved in their synthesis and breakdown constitute the "endocannabinoid system" (ECS). Although CB_1 and CB_2, to some extent together with the TRPV1 receptor, are phylogenetically and functionally unique [1,6], this is not the case for their ligands and associated enzymes. It is becoming increasingly clear how much the ECS *per se* is tightly intertwined with other signaling mechanisms. Some (if not all) of the "classical" endocannabinoids display "promiscuous" behavior by activating or blocking other receptors besides CB_1 or CB_2, with potencies that differ little from those with which they interact with *bona fide* CB receptors [1,7]. Furthermore, endocannabinoids have been found to belong to a large group of structurally related endogenous amides, esters, and ethers of fatty acids, which exist in a continuous dynamic equilibrium with each other. These molecules interact with a much wider range of receptors, including GPR55, GPR18, GPR119, TRPA1 (transient receptor potential ankyrin 1), TRPV1 (transient receptor potential channel type V1), other (TRP) cation channels, PPARs (peroxisome proliferator-activated receptors) as well as several nonreceptor targets [1,7–9]. Last but not the least, biochemical pathways for synthesis and degradation of endocannabinoids and their congeners show several crossroads with those of other bioactive lipids. This not only creates a number of regulatory nodes, but also results in the formation of "hybrid" structures, including prostamides and other oxidation products, which are often displaying bioactivity themselves (See also chapters 4 and 6 of this book) [10–13]. Taken together, there is a growing consensus that an "expanded" view of the ECS is more appropriate to study and understand its full dimensions [14,15]. In line with this, it has been suggested to introduce the term "endocannabinoidome" to describe this network of endocannabinoid-like mediators and their often-redundant metabolic enzymes and "promiscuous" targets [14]. Mediators belonging to the endocannabinoidome are fluctuating in a time and tissue-specific way, modulated by various endogenous (e.g., energy status, inflammation) and environmental factors, including diet [16–18]. Derived from fatty acids and able to interact with a plethora of molecular targets, these molecules are playing important roles as signaling molecules in the regulation of food-intake and energy homeostasis. It is conceivable that even subtle shifts in their relative concentrations can have biological implications, since different congeners possess different receptor affinities or intrinsic activities. The present chapter addresses some aspects of relation between dietary intake and the endocannabinoidome, focusing on the balance between *n*-6 and *n*-3 PUFAs, and its (possible) biological consequences. Until recently, the main focus was on endocannabinoids and related molecules synthesized from the most common fatty acids in higher animals, in particular arachidonic acid (AA) (20:4n–6), palmitic acid (16:0), oleic acid (18:1n–9), and stearic acid (18:0). Although the existence of n-3 PUFA-derived endocannabinoids has already been known for several years [19], their potential biological roles have only recently started to be explored. Dietary n-3 LC-PUFAs, in particular α-linolenic acid (ALA; 18:3n–3), eicosapentaenoic acid (EPA; 20:5n–3), docosapentaenoic acid (DPA; 22:5n–3), and docosahexaenoic acid (DHA; 22:6n–3) are receiving much interest because of their potential positive effects in health and disease. However, several questions remain, including on their long-term effects, doses, and mechanisms of action [5]. A number of recent studies suggest that some of the health effects of n-3 fatty acids may involve the endocannabinoidome.

2.2 MODULATION OF THE ENDOCANNABINOIDOME BY DIETARY FATTY ACIDS – BIOCHEMICAL ASPECTS

Most of the endocannabinoid-like molecules described so far have a fatty acid amide structure (lipid maps class FA08; http://www.lipidmaps.org). The "classical" endocannabinoid anandamide (N-arachidonoyletha-nolamine, AEA) is the best-known example of the N-acylethanolamine (NAE) subclass of fatty acid amides (Figure 2.1). Next to NAEs, several other fatty acid amides are found, including the primary fatty acid amides, the N-acylamino acids (= N-acylamines) and N-acylarylalkylamines (N-acyldopamines, N-acylserotonins) (Figure 2.1) [20,21].

Apparently, cells are able to "combine" different fatty acids and biogenic amines to make several possible permutations of different fatty acid amides [9,20]. In addition, studies suggest the formation of an obviously smaller number of different 2-acylglycerol esters, congeners of the endocannabinoid 2-arachidonoylglycerol (2-AG) (Chapter 1 of this book). The tissue concentration of several members of the endocannabinoidome was shown to fluctuate depending on the relative availability of their precursor fatty acids in the phospholipid membranes, which in turn is modulated by dietary supply and endogenous synthesis. As will be discussed in the following sections, a number of studies in rodents and humans support this concept by showing that increasing the relative proportion of n-3 LC-PUFAs in the diet can lead to a relative decrease in the formation of the AA (n-6)-derived "classical" endocannabinoids

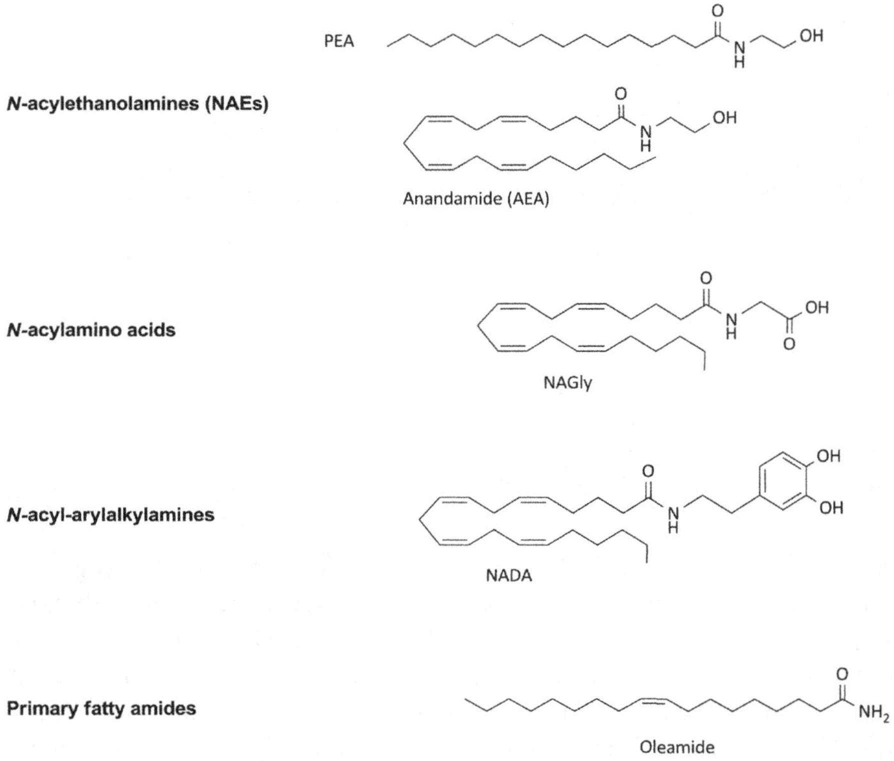

Fig. 2.1. Examples of different fatty acid amide structures.

AEA and 2-AG. At the same time, concentrations of *n*-3 fatty acid derived amides are found to increase with such diets [5,16,22,23]. A similar general principle applies to the local availability of amines such as ethanolamine, amino acids, or amine neurotransmitters, although it remains to be established to what extent this responds to dietary changes. For example, *N*-acyl dopamines have only been found in brain and dorsal root ganglia thus far [24,25]. Furthermore, we have previously shown that serotonin conjugates with fatty acids are predominantly present in the gut, where most of the body's serotonin resides [26]. Recently, the presence of *N*-arachidonoyl serotonin has also been demonstrated in bovine and human brain tissue [27]. As most, if not all, endocannabinoid-like molecules are rapidly broken down, the endocannabinoidome is considered a highly dynamic and versatile system to fine tune homeostasis. These features will also often limit the possibility to draw firm conclusions from

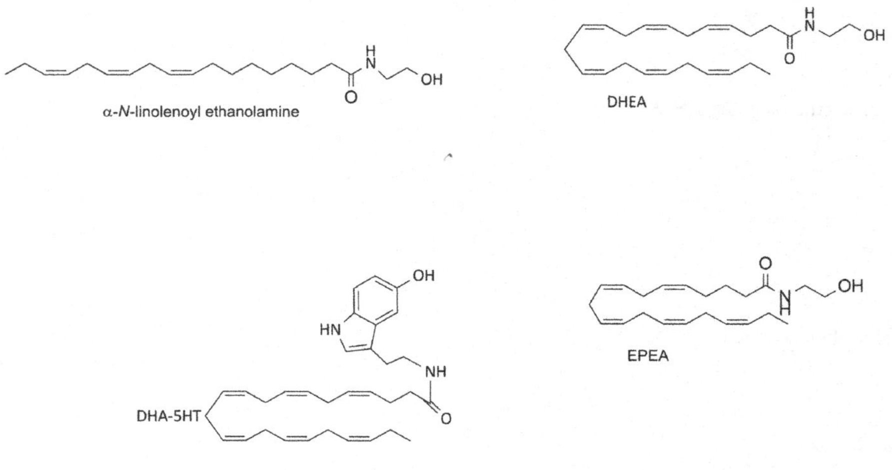

Fig. 2.2. *Structures of some n-3 PUFA-derived amides.*

plasma concentrations only and present great technical challenges in terms of bioanalysis and data-integration. *N*-Acyl conjugates of dietary important *n*-3 LC-PUFAs including DHA, EPA, DPA, and ALA with ethanolamine, dopamine, serotonin, and amino acids (see Figure 2.2 for some examples) have now been found in different organisms and tissues [5]. The presence of DHEA and EPEA was recently also demonstrated in cow's milk [28].

2.3 EFFECTS OF DIET – ANIMAL STUDIES

In 2001, Berger et al. [29] demonstrated that changing the lipid composition of a diet given to piglets causes differences in brain concentrations of fatty acid amides. A diet enriched with DHA given for 18 days increased both DHEA and EPEA levels in piglet brain. Since then similar studies (see Table 2.1 for overview), using different fatty acids and monitoring various fatty acid amide tissue concentrations have led to the concept of what may be called "natural combinatorial chemistry" in the formation of these molecules [9].

One of the most extensive studies in this respect so far was carried out by Artmann et al. [30], who fed diets with varying lipid enrichments, including a fish oil diet, to rats. After 1 week, different NAEs were measured in brain, liver, and jejunum. In brain, concentrations of most

Table 2.1 (Nonexhaustive) Overview of Dietary Studies in Animals Showing the Modulation of the Endocannabinoidome by n-3 Fatty Acids

References	Species	n-3 Lipid Source/Type	Study Duration	Analyzed NAEs	Analyzed Tissues	Principal Findings
Berger et al. [29]	Pigs (newborn)	DHA as triglycerides (in formula)	18 days	AEA, DHEA, and EPEA	Brain	Brain DHEA and EPEA ↑
Artmann et al. [30]	Sprague-Dawley rats	Fish oil	7 days	AEA, DHEA, EPEA OEA, PEA, and SEA	Brain, jejunum, and liver	Jejunum DHEA and EPEA ↑; no effect of fish oil on brain DHEA or EPEA
Batetta et al. [32]	Zucker fa/fa rats	Fish oil and krill oil	28 days	AEA	Heart, liver, and adipose tissue	Heart, liver, and adipose tissue AEA ↓
Wood et al. [31]	CD1 mice	Fish oil	14 days	AEA, DHEA, EPEA, OEA, and PEA	Brain, plasma	Brain AEA ↓; plasma OEA ↓; brain and plasma DHEA ↑. EPEA not detected?
Piscitelli et al. [33]	C57BL/6 mice	Krill oil (in high-fat diet)	56 days	AEA, OEA, and PEA	Adipose tissue, liver, muscle, kidney, heart	AEA in adipose tissue, muscle, kidney, heart ↓; OEA relatively stable
Balvers et al. [22,60]	C57BL/6 mice	Fish oil	42 days	AEA, DHEA, EPEA OEA, PEA, and SEA	Plasma, liver, ileum, and adipose tissue	Plasma, liver, ileum, and adipose tissue DHEA and EPEA ↑; EPEA only detectable in plasma after fish oil; increased levels after inflammatory stimulus
Rossmeisl et al. [35]	C57BL/6J mice	DHA and EPA as triglycerides or phospho-lipids, from fish oil	63 days	AEA, DHEA, EPEA	Adipose tissue	Adipose tissue AEA ↓; DHEA, and EPEA ↑. Negative correlation between adipose tissue DHEA levels and adipocyte size

NAEs, including DHEA did not change, which in retrospect may be due to the relatively short time of feeding the intervention diets. However, in liver and jejunum marked changes were observed, including increased DHEA and EPEA concentrations in the jejunum following the fish oil diet. In a similar study by Wood et al., a DHA-rich fish oil diet was fed for 2 weeks to mice, which resulted in increased plasma and brain DHEA levels. EPEA concentrations were not reported [31]. Interestingly, this study also showed that the fish oil diet decreased brain AEA and plasma OEA levels, which supports the concept of a compensatory decrease of the conjugation of fatty acids other than n-3 PUFAs. A similar observation was made in Zucker rats and in mice fed a high-fat diet with fish oil, showing a relative reduction of anandamide (AEA) levels in a variety of tissues, including heart, liver, adipose tissue, kidney, and muscle [32,33]. Collective data show that some NAEs (e.g., AEA and DHEA) are more sensitive to dietary modulation than others. An example of an NAE shown to be rather unresponsive to dietary modulation is palmitoylethanolamide (PEA), which is probably due to the fact that its precursor palmitoic acid is abundantly synthesized during normal fatty acid synthesis [23,34]. Recently, we showed in mice that a 6-week fish oil diet increased DHEA and EPEA concentrations in plasma, liver, ileum, and adipose tissue, which persisted during inflammatory conditions [22]. At the same time, tissue-specific reductions were observed for other NAEs, including AEA, OEA, and SEA. We did not find EPEA in plasma of mice on a standard diet, whereas this was the case when mice received a diet containing fish oil [22]. In a mouse study performed by Rossmeisl et al., n-3 fatty acids were given either as triglycerides via fish oil, or as phospholipids via krill oil [35]. This study showed that both forms reduced concentrations of AEA and increased those of DHEA and EPEA in adipose tissue. Interestingly, adipose tissue DHEA concentrations were higher after the phospholipid diet compared to the triglyceride group. Another example comes from the formation of N-acylserotonins in mouse intestinal tissue [26]. After a fish oil diet, concentrations of DHA- and EPA-serotonin increased. At the same time, concentrations of serotonin conjugates of arachidonic, stearic, and oleic acid decreased, whereas that of palmitic acid remained unchanged. Taken together, these findings support the concept of a competition mechanism between fatty acids for their conversion into fatty acid amides. Differences in amide patterns caused by (local) differences in the supply of fatty acids or amines are likely to be of physiological relevance. Results of these dietary interventions also illustrate the value of studying patterns of molecules instead of single compounds, and to include if possible different tissues. Furthermore, when biological effects of endocannabinoid-like molecules are investigated, a multiple component and multiple target strategy is preferable over a single-target approach.

2.4 EFFECTS OF DIET – HUMAN DATA

So far only a few studies have addressed the relationship between the intake or levels of specific fatty acids with patterns of fatty acid amides and (or) 2-acylglycerols in humans. In our laboratory, we observed significant correlations between plasma NAEs and their "precursor" serum-free fatty acid concentrations in humans during fasting and the postprandial state [36]. In a pilot study with three volunteers supporting analytical method development, we found that daily intake of 480 mg EPA plus 360 mg DHA for 3 weeks approximately doubled plasma DHEA levels. So far, we could not detect EPEA in human plasma. Two human intervention trials have examined the effect of taking krill oil containing high levels of *n*-3 fatty acids on (circulating) NAE/endocannabinoid levels [37,38]. The first study investigated the effect of a 4-week krill oil supplementation on circulating AEA and 2-AG concentrations in normo-weight, overweight, and obese subjects, mostly women [37]. 2-AG was reduced in obese persons after taking krill oil but no effects were found on AEA. In a more recent study, krill powder (= krill oil + proteins) was consumed by mildly obese men in which plasma concentrations of AEA, OEA, PEA, and 2-AG were measured after 12 and 24 weeks [38]. After 24 weeks, AEA, OEA, and PEA were significantly reduced compared to baseline levels, whereas 2-AG remained unaffected. The reductions in AEA and the other NAEs are in line with findings from animal studies as described above. Unfortunately, both human studies did not include DHEA and EPEA in their analysis.

2.5 *IN VITRO* FORMATION OF *N*-3 FATTY ACID DERIVED AMIDES

Using 3T3 F144 murine adipocytes, Matias et al. [39] demonstrated that incubation with DHA (100 μM for 72 h) reduced intracellular AEA concentrations, without affecting those of OEA and PEA. DHEA and EPEA were not part of the analytical panel. Using 3T3-L1 adipocytes we showed in our laboratory that incubation with 10 μM DHA or EPA for 24 h resulted in increased medium concentrations of DHEA and EPEA, respectively [40]. More recently, the formation of DHEA and EPEA was also demonstrated in prostate and breast cancer cell lines after incubation with DHA or EPA, respectively [41]. Furthermore, conversion of DHA to DHEA was observed in cultured hippocampal neurons [42] and in cultured neural stem cells (NSCs) under differentiating conditions [43]. Using porcine isolated intestinal segments we showed that increasing amounts of N-acyl serotonins were formed *ex vivo* after 1 h of incubation with a concentration series of serotonin (0, 3, 10, and 30 μM, respectively) [26].

2.6 BIOLOGICAL EFFECTS OF *N*-3 FATTY ACID DERIVED NAEs AND OTHER AMIDES

2.6.1 Receptor Interaction

Different receptor binding studies suggest that DHEA and EPEA are relatively weak ligands for CB receptors. In the 1990s, some studies compared binding affinity of DHEA to CB_1 receptors with that of anandamide [44,45]. Similarly, low affinity (compared to anandamide) binding of EPEA to CB_1 was shown [46]. More recently, Brown et al. reported values of 633 and 124 nM for binding of DHEA to mouse brain CB_1 receptors in the absence and presence of the fatty acid amide hydrolase (FAAH) inhibitor PMSF, respectively [4]. For EPEA binding to CB_1 (in the presence of PMSF), slightly lower K_i values were found. The same authors showed that DHEA and EPEA can bind to CB_2 receptors, albeit with lower affinities compared to those for CB_1. DHEA and EPEA behaved as CB_1 and CB_2 receptor agonists to mouse brain and CHO-hCB2 cell membranes as demonstrated by their ability to produce a concentration-dependent stimulation of $[^{35}S]GTP\gamma S$ binding [4]. In both membrane preparations, DHEA displayed higher potency than EPEA. Binding of DHEA to human CB_2 receptors with affinity in the nanomolar range was confirmed in our laboratory using a human CB_2 membrane preparation prepared from *Sf*9 cells. Data on the binding of DHEA, EPEA, or other *n*-3 derived amides to receptors other than CB_1 or CB_2 are not yet available. However, given their structural resemblance to, for example, AEA, there may be a number of candidates including, TRPV1. In addition, PPARγ receptors may play a role in the effects of DHEA/EPEA as will be described in subsequent sections.

2.6.2 Effects in Inflammation

Comparing of a series of structurally related NAEs (chain length C18 to C22; unsaturations 1*n* to 6*n*) we showed that DHEA was the most effective component in reducing nitric oxide (NO) release from stimulated RAW264.7 macrophages [47]. Potency of the NAEs in this assay correlated with chain length and unsaturation degree, displaying efficacy to inhibit NO release in the following order: DHEA > DEA > EPEA > AEA. NAEs with a chain length of 18 carbon atoms, including oleoylethanolamine (OEA) were not active. Interestingly, DHA, the *n*-3 fatty acid precursor of DHEA only induced a small albeit significant reduction of NO release in this assay. Nitric oxide produced by inducible NOS (iNOS) during pathological conditions is a late inflammatory mediator involved in apoptotic processes and pivotal in the immune response against pathogenic invaders. During chronic inflammatory conditions, prolonged NO production can be harmful, and limiting NO production has been shown to be beneficial in several animal models of disease [48–50]. Further studies [51] showed that DHEA also suppressed the early inflammatory mediator and chemoattractant CCL2 (MCP-1) and NO in LPS-stimulated mouse peritoneal macrophages. In RAW264.7 cells, CCL2 inhibition was found to take place at gene expression level and inhibition of NO release at the level of iNOS transcription. In differentiated 3T3-L1 adipocytes, DHEA and EPEA caused a reduction in LPS-induced CCL2 and IL-6 release [40]. In this assay, both LC-PUFA derived endocannabinoids were already effective at concentrations as low as 1 nM.

2.6.3 Effects on COX-2

To elucidate the underlying immunomodulatory mechanisms of DHEA, we investigated the involvement of different inflammatory mediators and transcription factors. NF-kB and IFN-β, both important players in the MyD88-dependent and the MyD88-independent pathway, respectively, turned out not to be involved in the DHEA-mediated effects in LPS-stimulated macrophages. However, DHEA dose-dependently reduced concentrations of the cyclooxygenase-2 (COX-2) metabolites PGD2, PGE_2, $PGF_{2\alpha}$, $TBXB_2$, and 12-HHTrE, whereas its precursor DHA was found to be inactive in this assay [51]. The mechanism behind this activity was found to involve a direct inhibition of enzyme activity combined with a slight reduction of enzyme expression, but not of its mRNA. COX-2 metabolizes AA into prostaglandins and thromboxanes, which exhibit specific regulatory functions in inflammatory processes [52]. Endocannabinoids such as AEA and 2-AG can also serve as substrates for

COX-2 thereby generating prostaglandin ethanolamides (PG-EA) and prostaglandin glycerol esters, respectively (see Chapters 4 and 6 of this book) [53–55]. It remains to be established whether DHEA serves as COX-2 inhibitor or as substrate leading to metabolites with lower pro-inflammatory activity.

2.6.4 Potential Roles of DHEA and EPEA as Endogenous Mediators of Inflammation

Besides diet, other factors, including inflammatory status can influence the levels of components of the endocannabinoidome in a time- and tissue-specific way. Several studies are now supporting the viewpoint that both AEA and 2-AG are playing pivotal roles in the modulation of inflammatory processes and in reestablishing homeostasis after an inflammatory event [3,56–58]. Many immune cells are able to release 2-AG, AEA, and other endocannabinoids-like mediators, and their levels are often found to be increased in animal disease models and in certain diseases in humans [3,58–60]. Recent studies indicated that the same might be true for n-3 PUFA-derived endocannabinoids. For example, DHEA and EPEA tissue concentrations were found to increase after an inflammatory stimulus in mice fed fish oil [22,61]. These animals also showed reduced concentrations of COX-2-derived eicosanoids including PGD_2, PGE2, and their metabolites 13,14-dihydro-15-keto-PGD_2 and –PGE_2, $PGF_{2\alpha}$ $TBXB_2$. Together, this suggests that n-3 fatty acid derived amides might play a role as endogenous anti-inflammatory mediators.

2.6.5 Effects in Tumor Cell Lines

DHEA and EPEA were shown to possess potential anticarcinogenic properties in prostate and breast cancer cell lines [4,8]. Brown et al. [4] reported that DHEA and EPEA displayed antiproliferative and cell growth inhibitory effects in LNCaP and PC3 human prostate cancer cell lines. Remarkably, both NAEs showed greater antiproliferative potency than their parent compounds DHA and EPA. The inhibition resulted from an increased apoptosis and changes in cell cycle arrest. The CB_1- and CB_2-selective antagonists, AM281 and AM630, separately or in combination, reduced the antiproliferative effect of EPEA in PC3 cells but not in LNCaP cells, while they potentiated the effect of DHEA in both cell lines. Rovito et al. [62] showed that DHEA and EPEA specifically reduced cell viability in a MCF-7 human breast cancer cell line. This effect resulted from the induction of autophagic cell death. Activation of PPARγ was found to play a crucial role. The PPARγ receptor antagonist GW9662 prevented the autophagic process induced by the treatment with DHEA or EPEA. Moreover, the inhibitory effect of *n*-3 NAEs on cell growth was potentiated by the PPARγ ligand BRL49653 and blocked by GW9662.

2.6.6 Role in Neuroprotection and Neurogenesis

It has been suggested that DHEA and other DHA conjugates are important for brain development and the maintenance of brain functioning, and that they play roles in neuroprotection and the control of inflammation during disease or resulting from tissue damage. Like their parent compound DHA, DHEA, and other conjugates are found in relatively high concentrations in brain [29,31,45,63]. Although several other pathways are being proposed to explain the effects of DHA on brain [64] it is well conceivable that mechanisms taking place via their amine conjugates will be involved here as well. For example, the presence of DHEA (synaptamide as it was called by the authors) was demonstrated in mouse hippocampus and shown to be a potent stimulator of neurite growth and synaptogenesis in cultured hippocampal neurons [42,65,66]. Furthermore, it enhanced glutamatergic synaptic activity. Again, the bioactivities of DHEA were higher than those of the parent compound DHA. The effects of DHEA on hippocampal neuronal development were found to be not affected by CB_1 and CB2 agonists or antagonists [65]. DHEA was also found to potently induce neuronal differentiation in cultured NSCs [43]. Treatment with nanomolar concentrations of DHEA significantly increased the number of microtubule-associated protein 2 (MAP2) and neuron-specific class III beta-tubulin (Tuj-1)-positive neurons and their corresponding protein levels. The neurogenic property of DHEA was found to be mediated through protein kinase A (PKA)/cAMP response element binding protein (CREB) signaling pathway. PKA inhibitors or PKA knockdown was able to abolish the DHEA-induced neuronal differentiation of NSCs. A series of oxygenated metabolites from DHEA were identified in mice brain that regulated leukocyte motility [67,68]. The authors conclude that these metabolites might serve as anti-inflammatory and organ-protective mediators in brain.

2.7 CONCLUSIONS AND FUTURE PERSPECTIVES

It is obvious that the endocannabinoidome is modulated by various (patho-)physiologic and environmental factors in a time- and tissue-specific manner. It holds many promises for new "food" and "pharma" applications, as it is crucially involved in many processes and disorders. However, understanding or modulating the endocannabinoidome requires that its subtle and versatile character is properly addressed. The link with dietary fatty acids as discussed in the present chapter is just one aspect of this. Accumulating data suggest that *n*-3 PUFA-derived endocannabinoids-like mediators may play important roles in the suggested health effects of their parent fatty acids. A schematic overview of the bioactivities of DHEA (and EPEA) is given in Figure 2.3.

Future bioactivity studies *in vivo*, if possible also in humans and preferably using combinations of compounds and taking into account the system dynamics are merited. In case of fatty acid amides, it is also of interest to pay more attention to the formation and bioactivity of conjugates of other amines, in particular, amino acids and neurotransmitters.

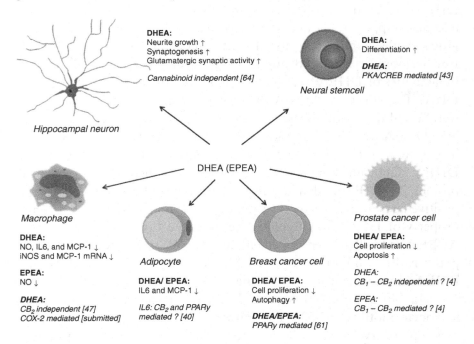

Fig. 2.3. Schematic overview of bioactivities found for DHEA so far, with [references].

REFERENCES

[1] Pertwee RG, Howlett AC, Abood ME, et al. International Union of Basic and Clinical Pharmacology. LXXIX. Cannabinoid receptors and their ligands: beyond CB1 and CB2. Pharmacol Rev 2010;62(4):588–631.

[2] Pertwee RG, Howlett AC, Abood M, et al. IUPHAR database (IUPHAR-DB). Cannabinoid receptors, introductory chapter 2011. Last modified on 07/12/2011. Last Accessed on 28/02/2014; Available from: http://www.iuphar-db.org/DATABASE/FamilyIntroductionForw ard?familyId=13.

[3] Witkamp R, Meijerink J. The endocannabinoid system: an emerging key player in inflammation. Curr Opin Clin Nutr Metab Care 2014;17(2):130–8.

[4] Brown I, Cascio MG, Wahle KW, et al. Cannabinoid receptor-dependent and -independent anti-proliferative effects of omega-3 ethanolamides in androgen receptor-positive and -negative prostate cancer cell lines. Carcinogenesis 2010;31(9):1584–91.

[5] Meijerink J, Balvers M, Witkamp R. *N*-acyl amines of docosahexaenoic acid and other *n*-3 polyunsaturated fatty acids – from fishy endocannabinoids to potential leads. Br J Pharmacol 2013;169(4):772–83.

[6] Elphick MR. The evolution and comparative neurobiology of endocannabinoid signalling. Philos Trans R Soc Lond B Biol Sci 2012;367(1607):3201–15.

[7] Alexander SPH, Kendall DA. The complications of promiscuity: endocannabinoid action and metabolism. Br J Pharmacol 2007;152(5):602–23.

[8] Brown I, Cascio MG, Rotondo D, et al. Cannabinoids and omega-3/6 endocannabinoids as cell death and anticancer modulators. Prog Lipid Res 2013;52(1):80–109.

[9] Di Marzo V, Bisogno T, De Petrocellis L. Endocannabinoids and related compounds: walking back and forth between plant natural products and animal physiology. Chem Biol 2007;14(7):741–56.

[10] Silvestri C, Martella A, Poloso NJ, et al. Anandamide-derived prostamide F2α negatively regulates adipogenesis. J Biol Chem 2013;288(32):23307–21.

[11] Woodward DF, Jones RL, Narumiya S. International Union of Basic and Clinical Pharmacology. LXXXIII: Classification of prostanoid receptors, updating 15 years of progress. Pharmacol Rev 2011;63(3):471–538.

[12] Woodward DF, Liang Y, Krauss AHP. Prostamides (prostaglandin-ethanolamides) and their pharmacology. Br J Pharmacol 2007;153(3):410–9.

[13] Woodward DF, Wang JW, Poloso NJ. Recent progress in prostaglandin F2α ethanolamide (prostamide F2α) research and therapeutics. Pharmacol Rev 2013;65(4):1135–47.

[14] Maione S, Costa B, Di Marzo V. Endocannabinoids: a unique opportunity to develop multi-target analgesics. Pain 2013;154(Suppl. 1):S87–93.

[15] Silvestri C, Di Marzo V. The endocannabinoid system in energy homeostasis and the etiopathology of metabolic disorders. Cell Metab 2013;17(4):475–90.

[16] Maccarrone M, Gasperi V, Catani MV, et al. The endocannabinoid system and its relevance for nutrition. Annu Rev Nutr 2010;30(1):423–40.

[17] Hansen HS, Artmann A. Endocannabinoids and nutrition. J Neuroendocrinol 2008;20(s1): 94–9.

[18] Hansen HS, Diep TA. *N*-Acylethanolamines, anandamide and food intake. Biochem Pharmacol 2009;78(6):553–60.

[19] Bisogno T, Delton-Vandenbroucke I, Milone A, et al. Biosynthesis and inactivation of *N*-arachidonoylethanolamine (anandamide) and *N*-docosahexaenoylethanolamine in bovine retina. Arch Biochem Biophys 1999;370(2):300–7.

[20] Bradshaw HB, Raboune S, Hollis JL. Opportunistic activation of TRP receptors by endogenous lipids: exploiting lipidomics to understand TRP receptor cellular communication. Life Sci 2013;92(8–9):404–9.

[21] Farrell EK, Merkler DJ. Biosynthesis, degradation and pharmacological importance of the fatty acid amides. Drug Discov Today 2008;13(13–14):558–68.

[22] Balvers MJ, Verhoeckx KM, Bijlsma S, et al. Fish oil and inflammatory status alter the n-3 to n-6 balance of the endocannabinoid and oxylipin metabolomes in mouse plasma and tissues. Metabolomics 2012;8(6):1130–47.

[23] Hansen HS. Effect of diet on tissue levels of palmitoylethanolamide. CNS Neurol Disord Drug Targets 2013;12(1):17–25.

[24] Chu CJ, Huang SM, De Petrocellis L, et al. N-Oleoyldopamine, a novel endogenous capsaicin-like lipid that produces hyperalgesia. J Biol Chem 2003;278(16):13633–9.

[25] Hu SSJ, Bradshaw HB, Benton VM, et al. The biosynthesis of N-arachidonoyl dopamine (NADA), a putative endocannabinoid and endovanilloid, via conjugation of arachidonic acid with dopamine. Prostaglandins Leukot Essent Fatty Acids 2009;81(4):291–301.

[26] Verhoeckx KCM, Voortman T, Balvers MGJ, et al. Presence, formation and putative biological activities of N-acyl serotonins, a novel class of fatty-acid derived mediators, in the intestinal tract. Biochim Biophys Acta 2011;1811(10):578–86.

[27] Siller M, Goyal S, Yoshimoto FK, et al. Oxidation of endogenous N-arachidonoylserotonin by human cytochrome P450 2U1. J Biol Chem 2014;.

[28] Gouveia-Figueira S, Nording ML. Development and validation of a sensitive UPLC-ESI-MS/MS method for the simultaneous quantification of 15 endocannabinoids and related compounds in milk and other biofluids. Anal Chem 2013;86(2):1186–95.

[29] Berger A, Crozier G, Bisogno T, et al. Anandamide and diet: inclusion of dietary arachidonate and docosahexaenoate leads to increased brain levels of the corresponding N-acylethanolamines in piglets. Proc Natl Acad Sci USA 2001;98(11):6402–6.

[30] Artmann A, Petersen G, Hellgren LI, et al. Influence of dietary fatty acids on endocannabinoid and N-acylethanolamine levels in rat brain, liver and small intestine. Biochim Biophys Acta 2008;1781(4):200–12.

[31] Wood JT, Williams JS, Pandarinathan L, et al. Dietary docosahexaenoic acid supplementation alters select physiological endocannabinoid-system metabolites in brain and plasma. J Lipid Res 2010;51(6):1416–23.

[32] Batetta B, Griinari M, Carta G, et al. Endocannabinoids may mediate the ability of (n-3) fatty acids to reduce ectopic fat and inflammatory mediators in obese Zucker rats. J Nutr 2009;139(8):1495–501.

[33] Piscitelli F, Carta G, Bisogno T, et al. Effect of dietary krill oil supplementation on the endocannabinoidome of metabolically relevant tissues from high fat-fed mice. Nutr Metab 2011;8(1):51.

[34] Balvers MG, Verhoeckx KC, Meijerink J, et al. Measurement of palmitoylethanolamide and other N-acylethanolamines during physiological and pathological conditions. CNS Neurol Disord Drug Targets 2013;12(1):23–33.

[35] Rossmeisl M, Macek Jilkova Z, Kuda O, et al. Metabolic effects of n-3 PUFA as phospholipids are superior to triglycerides in mice fed a high-fat diet: possible role of endocannabinoids. PLoS ONE 2012;7(6):e38834.

[36] Joosten MM, Balvers MG, Verhoeckx KC, et al. Plasma anandamide and other N-acylethanolamines are correlated with their corresponding free fatty acid levels under both fasting and non-fasting conditions in women. Nutr Metab (Lond) 2010;7:49.

[37] Banni S, Carta G, Murru E, et al. Krill oil significantly decreases 2-arachidonoylglycerol plasma levels in obese subjects. Nutr Metab 2011;8.

[38] Berge K, Piscitelli F, Hoem N, et al. Chronic treatment with krill powder reduces plasma triglyceride and anandamide levels in mildly obese men. Lipids Health Dis 2013;12(1):78.

[39] Matias I, Carta G, Murru E, et al. Effect of polyunsaturated fatty acids on endocannabinoid and *N*-acyl-ethanolamine levels in mouse adipocytes. Biochim Biophys Acta 2008;1781 (1–2):52–60.

[40] Balvers MGJ, Verhoeckx KCM, Plastina P, et al. Docosahexaenoic acid and eicosapentaenoic acid are converted by 3T3-L1 adipocytes to *N*-acyl ethanolamines with anti-inflammatory properties. Biochim Biophys Acta 2010;1801(10):1107–14.

[41] Brown I, Wahle KWJ, Cascio MG, et al. Omega-3 *N*-acylethanolamines are endogenously synthesised from omega-3 fatty acids in different human prostate and breast cancer cell lines. Prostaglandins Leukot Essent Fatty Acids 2011;85(6):305–10.

[42] Kim HY, Moon HS, Cao D, et al. *N*-Docosahexaenoylethanolamide promotes development of hippocampal neurons. Biochem J 2011;435(2):327–36.

[43] Rashid MA, Katakura M, Kharebava G, et al. *N*-Docosahexaenoylethanolamine is a potent neurogenic factor for neural stem cell differentiation. J Neurochem 2013;125(6):869–84.

[44] Felder CC, Briley EM, Axelrod J, et al. Anandamide, an endogenous cannabimimetic eicosanoid, binds to the cloned human cannabinoid receptor and stimulates receptor-mediated signal transduction. Proc Natl Acad Sci 1993;90(16):7656–60.

[45] Sheskin T, Hanus L, Slager J, et al. Structural requirements for binding of anandamide-type compounds to the brain cannabinoid receptor. J Med Chem 1997;40(5):659–67.

[46] Adams IB, Ryan W, Singer M, et al. Evaluation of cannabinoid receptor binding and *in vivo* activities for anandamide analogs. J Pharmacol Exp Ther 1995;273(3):1172–81.

[47] Meijerink J, Plastina P, Vincken J-P, et al. The ethanolamide metabolite of DHA, docosahexaenoylethanolamine, shows immunomodulating effects in mouse peritoneal and RAW264.7 macrophages: evidence for a new link between fish oil and inflammation. Br J Nutr. 2011;105(12):1798–807.

[48] Detmers PA, Hernandez M, Mudgett J, et al. Deficiency in inducible nitric oxide synthase results in reduced atherosclerosis in apolipoprotein E-deficient mice. J Immunol 2000;165(6):3430–5.

[49] Charbonneau A, Marette A. Inducible nitric oxide synthase induction underlies lipid-induced hepatic insulin resistance in mice: potential role of tyrosine nitration of insulin signaling proteins. Diabetes 2010;59(4):861–71.

[50] Pilon G, Charbonneau A, White PJ, et al. Endotoxin mediated-iNOS induction causes insulin resistance via ONOO-induced tyrosine nitration of IRS-1 in skeletal muscle. PLoS One 2010;5(12).

[51] Meijerink J, Poland M, Balvers MG, et al. Inhibition of COX-2-mediated eicosanoid production plays a major role in the anti-inflammatory effects of the endocannabinoid N-docosahexaenoylethanolamine (DHEA) in macrophages. Br J Pharmacol 2014; DOI: 10.1111/bph.12747.

[52] Calder PC. Polyunsaturated fatty acids and inflammatory processes: new twists in an old tale. Biochimie 2009;91(6):791–5.

[53] Yu M, Ives D, Ramesha CS. Synthesis of prostaglandin E2 ethanolamide from anandamide by cyclooxygenase-2. J Biol Chem 1997;272(34):21181–6.

[54] Kozak KR, Crews BC, Morrow JD, et al. 15-Lipoxygenase metabolism of 2-arachidonylglycerol. Generation of a peroxisome proliferator-activated receptor alpha agonist. J Biol Chem 2002;277:23278–86.

[55] Rouzer CA, Marnett LJ. Endocannabinoid oxygenation by cyclooxygenases, lipoxygenases, and cytochromes P450: cross-talk between the eicosanoid and endocannabinoid signaling pathways. Chem Rev 2011;111(10):5899–921.

[56] Klein TW. Cannabinoid-based drugs as anti-inflammatory therapeutics. Nat Rev Immunol 2005;5(5):400–11.

[57] Miller AM, Stella N. CB2 receptor-mediated migration of immune cells: it can go either way. Br J Pharmacol 2008;153(2):299–308.

[58] Alhouayek M, Masquelier J, Muccioli GG. Controlling 2-arachidonoylglycerol metabolism as an anti-inflammatory strategy. Drug Discov Today 2013.

[59] Alhouayek M, Muccioli GG. The endocannabinoid system in inflammatory bowel diseases: from pathophysiology to therapeutic opportunity. Trends Mol Med 2012;18(10):615–25.

[60] Scotter EL, Abood ME, Glass M. The endocannabinoid system as a target for the treatment of neurodegenerative disease. Br J Pharmacol 2010;160(3):480–98.

[61] Balvers MGJ, Verhoeckx KCM, Meijerink J, et al. Time-dependent effect of in vivo inflammation on eicosanoid and endocannabinoid levels in plasma, liver, ileum and adipose tissue in C57BL/6 mice fed a fish-oil diet. Int Immunopharmacol 2012;13(2):204–14.

[62] Rovito D, Giordano C, Vizza D, et al. Omega-3 PUFA ethanolamides DHEA and EPEA induce autophagy through PPARγ activation in MCF-7 breast cancer cells. J Cell Physiol 2013;228(6):1314–22.

[63] Tan B, O'Dell DK, Yu YW, et al. Identification of endogenous acyl amino acids based on a targeted lipidomics approach. J Lipid Res 2010;51(1):112–9.

[64] Bazan NG, Molina MF, Gordon WC. Docosahexaenoic acid signalolipidomics in nutrition: significance in aging, neuroinflammation, macular degeneration, Alzheimer's, and other neurodegenerative diseases. Annu Rev Nutr 2011;31(1):321–51.

[65] Kim HY, Spector AA. Synaptamide, endocannabinoid-like derivative of docosahexaenoic acid with cannabinoid-independent function. Prostaglandins Leukot Essent Fatty Acids 2013;88(1):121–5.

[66] Kim H-Y, Spector AA, Xiong Z-M. A synaptogenic amide N-docosahexaenoylethanolamide promotes hippocampal development. Prostaglandins Other Lipid Mediators 2011;96(1–4):114–20.

[67] Yang R, Fredman G, Krishnamoorthy S, et al. Decoding functional metabolomics with docosahexaenoyl ethanolamide (DHEA) identifies novel bioactive signals. J Biol Chem 2011;286(36):31532–41.

[68] Shinohara M, Mirakaj V, Serhan CN. Functional metabolomics reveals novel active products in the DHA metabolome. Front Immunol 2012;3:1–9.

CHAPTER 3

N-Acyl Amides: Ubiquitous Endogenous Cannabimimetic Lipids That Are in the Right Place at the Right Time

Emma Leishman, Heather B. Bradshaw

3.1 *N*-ACYL AMIDES: ALL IN THE FAMILY

Ananadmide, *N*-arachidonoyl ethanolamide (AEA) was the first identi-fied endogenous cannabinoid and belongs to a structural group of com-pounds (hereafter referred to as a "family") of *N*-acyl ethanolamines (NAEs). Whereas AEA is formed from arachidonic acid, other NAEs are formed from the conjugation of other fatty acids and ethanolamine (Figure 3.1). Nearly all endogenous fatty acids have been measured as ethanolamide conjugates in mammalian tissues [1]. In fact, the other NAEs are more abundant than AEA in the mammalian brain [2]. NAE

The Endocannabinoidome: The World of Endocannabinoids and Related Mediators. DOI: 10.1016/B978-0-12-420126-2.00003-1
Copyright © 2015 Elsevier Inc. All rights reserved

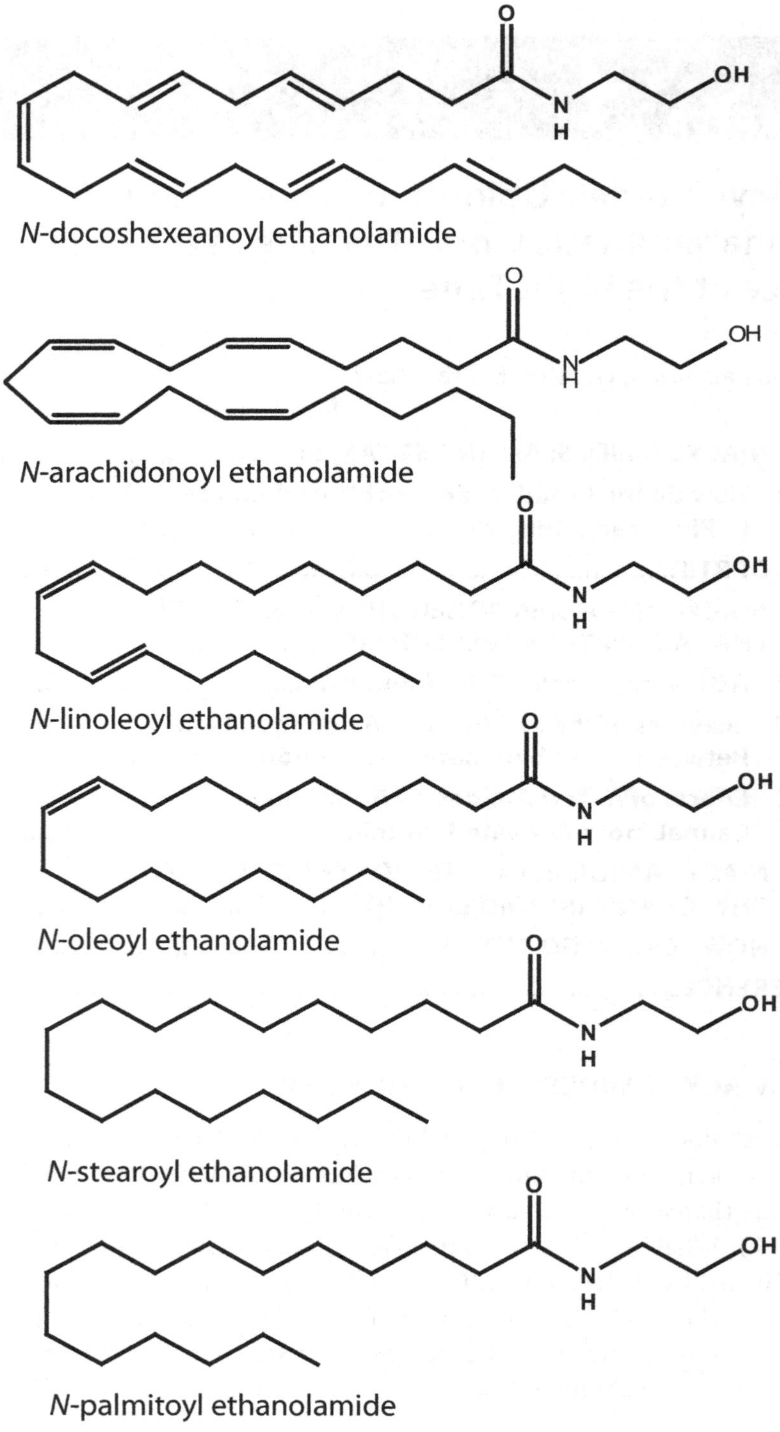

N-docoshexeanoyl ethanolamide

N-arachidonoyl ethanolamide

N-linoleoyl ethanolamide

N-oleoyl ethanolamide

N-stearoyl ethanolamide

N-palmitoyl ethanolamide

Fig. 3.1. Structures of endogenous N-acyl ethanolamides.

are not, however, unique to mammals in that the first to be identified was *N*-palmitoyl ethanolamine (PEA), the conjugate of palmitic acid and ethanolamine was found in soybeans, peanuts, and egg yolks where it was further determined that PEA had anti-inflammatory effects [3]. PEA has been measured in a wide variety of tissues including the rodent brain, liver, and skeletal muscle [4]. PEA selectively activates CB_2 over CB_1 [5], however, it is not clear that it is a primary ligand at this receptor. *N*-Linoleoyl ethanolamine (LEA) has also been detected in mouse and pig brain extracts and is produced in murine macrophage, leukocyte, and neuroblastoma cells and thought to be analogous to AEA because it can activate CB receptors, albeit at a much lower efficacy than AEA [6]. We recently showed that LEA, PEA, and *N*-oleoyl ethanolamine (OEA) are also present in Drosophila [7].

NAEs are, however, only one family in the much broader and diverse group of *N*-acyl amides. Structurally, *N*-acyl amides consist of a fatty acid (*acyl* group on chemical parlance) conjugated through and amide bond to a simple amine, such as ethanolamine in the case of the NAEs. Figure 3.2 represents the simple structure of this class of compounds. In place of the ethanolamine group, amines such as amino acids or amino acid derivatives can conjugate with fatty acids to create *N*-acyl amides. It is unknown how many *N*-acyl amides exist in nature. If we provide a theoretical construct that if you consider seven common fatty acids in the mammalian systems (arachidonic, stearic, docosahexaenoic, oleic, palmitic, linolenic, and linoleic) and conjugate them to 20 amino acids and 4 common amines (ethanolamine, dopamine, gamma-amino butyric

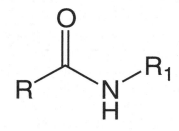

R = fatty acid, R_1 = amine

Fig. 3.2. Generic structure of an N-acyl amide.

acid (GABA), and taurine), there are 168 possible N-acyl amide combinations that could be formed in mammals. It must be noted that some of these combinations are still theoretical in that the standards have not been synthesized and of those that have, some lipids have not yet been identified in biological systems. Conceptually, these N-acyl amides that have been identified are considered "orphan lipids" that have simply not been matched to a receptor is a working hypothesis we and others have been exploring for many years [8].

3.1.1 How do the Broader Family of *N*-Acyl Amides Relate to Phytocannabinoids?

Although there are over 80 identified cannabinoids in the plant *Cannabis sativa*, Δ^9-tetrahydrocannabinol (THC) is hypothesized to be responsible for most of the psychoactive effects of marijuana [5]. How many of these additional phytocannabinoids act at molecular level are addressed in other chapters of this book, so our aim here is to remind the reader that while THC is thus far the most studied and understood cannabinoid, there are multiple receptor targets for phytocannabinoids [9,10]. It is important to discuss N-acyl amide activity in the context of cannabinoids, we have provided a general summary of a selection of phytocannabinoids and their corresponding receptor targets in Table 3.1. Importantly, Table 3.1 shows that many of these additional phytocannabinoids activate TRP receptors and that there is a considerable amount of "cross-reactivity" among ligands and receptors.

At present, there are two officially recognized cannabinoid receptors, CB_1 and CB_2, although other receptors such as GPR18, GPR55, and the TRPV receptors are being recognized as part of the larger cannabinoid system [11–14]. Although there are some endogenous ligands (endocannabinoid; eCBs) for the recognized and these putative phytocannabinoid receptors, since there are only two officially recognized cannabinoid receptors, the current number of eCBs is limited. If eCBs were instead defined as ligands for any receptor for which a phytocannabinoid binds with efficacy, then there would be many more endogenous lipid signaling molecules classified as eCBs; however, that discussion is also beyond the scope of this chapter. Suffice to say that N-acyl amides present a novel avenue for the expansion of both putative eCBs as well as a means to understand how phytocannabinoids are working in the body. Here, we

Table 3.1 Summary Table of Phytocannabinoid Effects on Receptors that have Endogenous N-acyl Amide Ligands

Phytocannabinoid	Receptor Targets
Δ^9-tetrahydrocannabinol (THC)	CB_1 agonist [1] CB_2 agonist [1] GPR18 agonist [3] TRPV2 agonist [4]
Cannabidiol (CBD)	CB_1 indirect antagonist [5] CB_2 indirect antagonist [5] GPR18 antagonist [3] TRPA1 agonist [4] TRPV1 agonist [6] TRPV2 agonist [7] TRPV3 agonist [8]
Cannabinol (CBN)	Very weak CB_1 agonist [1] Very weak CB_2 agonist [1] TRPV4 antagonist [8]
Cannabigerol (CBG)	CB_1 competitive antagonist [9] TRPA1 agonist [10] TRPV1 agonist [11] TRPV2 agonist [4] TRPV4 antagonist [8]
Tetrahydrocannabivarin (THCV)	CB_1 antagonist [12] TRPV1 agonist [4] TRPV2 agonist [4] TRPV3 agonist [8] TRPV4 agonist [8]
Cannabidivarin (CBDV)	TRPV2 agonist [4] TRPV4 agonist [8]
Cannabigerovarin (CBGV)	TRPV1 agonist [4] TRPV2 agonist [4] TRPV3 antagonist [8] TRPV4 antagonist [8]
Cannabigerolic acid (CBGA)	TRPV3 antagonist [8]

[1] Felder et al. (1995); [2] Ryberg et al. (2007); [3] McHugh et al. (2012); [4] De Petrocellis et al. (2011); [5] Pertwee (2008); [6] Bisogno et al. (2001); [7] Qin et al. (2008); [8] De Petrocellis et al. (2012); [9] Cascio (2010); [10] De Petrocellis et al. (2008); [11] Ligresti et al. (2006); [12] Pertwee et al. (2007).

will summarize what is known about activity of *N*-acyl amides at the receptors highlighted in Table 3.1 that are activated by phytocannabinoids and have summarized the *N*-acyl amides that activate these same receptors in Table 3.2. Given that both these class of receptors and ligands do not officially belong to a pharmacological class; however, they have direct relationships to those receptor and ligands that are associated with the recognized cannabinoids, they will hereafter be referred to simply as "Cannabimimetic."

Table 3.2 *N*-Acyl Amides that Activate Cannabimimetic Receptors	
Receptor	Ligand
CB₁	AEA [1] *N*-Arachidonoyl dopamine [3] LEA [4]
CB₂	AEA [1] PEA [5] LEA [4]
GPR18	AEA [8] *N*-Arachidonoyl glycine [8] *N*-Arachidonoyl serine [7]
TRPV1	AEA [12] *N*-Arachidonoyl dopamine [13] *N*-Oleoyl dopamine [14] *N*-Arachidonoyl taurine [15] *N*-Docosahexaenoyl ethanolamine [16] *N*-Docosahexaenoyl GABA [16] *N*-Docosahexaenoyl aspartic acid [16] *N*-Docosahexaenoyl glycine [16] *N*-Docosahexaenoyl serine [16] *N*-Arachidonoyl GABA [16] *N*-Linoleyl GABA [16]
TRPV2	*N*-Acyl proline mixture [16] *N*-Palmitoyl tyrosine [16]
TRPV3 (ligands acted as *antagonists*)	*N*-Docosahexaenoyl valine [16] *N*-Linoleoyl valine [16] *N*-Oleoyl valine [16] *N*-Stearoyl valine [16]
TRPV4	*N*-Arachidonoyl taurine [15] *N*-Arachidonoyl tyrosine [16] *N*-Linoleoyl tyrosine [16] *N*-Palmitoyl tyrosine [16] *N*-Docosahexaenoyl tryptophan [16] *N*-Arachidonoyl tryptophan [16] *N*-Linoleoyl tryptophan [16]

[1] Devane et al. (1992); [2] Sugiura et al. (1995); [3] Bisogno et al. (2000); [4] Lin et al. (1998); [5] Jaggar et al. (1998); [6] Jarai et al. (1999); [7] McHugh et al. (2012); [8] McHugh et al. (2010); [9] Lauckner et al. (2008); [10] Oka et al. (2007); [11] Oka et al. (2009); [12] Zymunt et al. (1999); [13] Hu et al. (2009); [14] Chu et al. (2003); [15] Saghatelian et al. (2006); [16] Raboune et al. (2014).

3.2 GPR18

"Endothelial anandamide receptor" was so named due to its effect in the endothelial cells lining vasculature that responded to AEA resulting in vasodilation in a mechanism independent of CB_1 or CB_2 [15]. The effects of AEA were mimicked by abnormal-cannabidiol (Abn-CBD/O-1918), a synthetic cannabinoid analog of CBD, which led to the renaming of the receptor as the "Abn-CBD receptor." Over the past 7 years, our group has provided evidence that Abn-CBD receptor and GPR18 were one and the same. *N*-Acyl amide, *N*-arachidonoyl glycine (NAGly), is a potent agonist at GPR18 and drives migration in both microglial and human endometrial carcinoma cells through GPR18-dependent pathways [12,16]. NAGly is formed from AEA via two distinct pathways that can both occur in mammalian cells. The first is a direct oxidation of AEA by alcohol dehydrogenase (ADH), and the second is via the conjugation of arachidonic acid and glycine in a FAAH-dependent reaction [17]. Unlike AEA, NAGly has no affinity for cannabinoid receptors or TRPV1 [18]. However, NAGly possesses antinociceptive and antiinflammatory properties in several animal models of pain in a similar manner to AEA [19].

3.3 *N*-ACYL AMIDES/ENDOGENOUS CANNABINOIDS THAT ACTIVATE TRP RECEPTORS

Transient receptor potentials (TRPs) regulate cation entry to affect numerous intracellular signaling pathways and are being considered the "ionotropic cannabinoid" receptors [20]. Indeed, phytocannabinoids can activate TRPV1, 2, 3, and 4 [21–25]. One of the more thoroughly studied TRPs is TRPV1 (vanilloid 1), which is extensively involved in nociception and sensory transmission. TRPV1 can open in response to exogenous ligands, such as capsaicin, and to changes in physical and chemical environments, such as heat or acidic pH. The binding of endogenous lipids can modify the response of TRPV1 to exogenous challenges [8]. Although mostly studied in the dorsal root ganglion of the spinal cord, TRPV1 can also be found in the brain and spinal cord [26]. AEA and many of its direct or indirect metabolites are agonists at TRPV1 [27–33]. Zygmunt et al. first characterized AEA as a TRPV1 agonist. They found that the vasodilatory effects of AEA couldn't be blocked by CB_1 receptor antagonists, but could be blocked by capsazepine, a TRPV1 receptor antagonist [34]. AEA activates TRPV1 through a PKC-dependent pathway, which leads to the perception of a painful stimulus. TRPV1 then desensitizes, which is thought to be a mediator of AEA's analgesic effects outside of the cannabinoid receptors [4].

N-Arachidonoyl dopamine, NADA, in addition to activating CB_1, is an agonist at TRPV1. This could potentially explain the paradoxical hyperalgesic effects of NADA administrations in rodent models [35]. The hyperalgesia seen was behaviorally similar to that produced by capsaicin, a potent TRPV1 agonist [36]. N-Oleoyl dopamine (OLDA) is a structural analog of NADA that lacks any functional affinity for classical cannabinoid receptors. However, OLDA caused calcium influx in TRPV1 transfected HEK cells at an EC_{50} of 36 nM, making it an agonist at TRPV1 [37]. The first noncannabinoid N-acyl amide to be assigned a receptor and biological significance was N-arachidonoyl taurine, which activates TRPV1 and TRPV4 receptors at micromolar concentrations [38].

Given that certain N-acyl taurines and additional N-arachidonyl amides were TRPV agonists, the Bradshaw group used calcium imaging to challenge cells expressing TRPV1-4 receptors with 81 different N-acyl amides. At TRPV1, three N-acyl GABAs, two novel NAEs, N-docosahexaenoyl aspartic acid, N-docosahexaenoyl glycine, and N-docosahexaenoyl serine all demonstrated agonist activity at low micromolar concentrations [39]. Hinting at the presence of an ensemble effect, combining all the individual agonists in a single assay produced an EC_{50} value in the nanomolar range (unpublished data). At TRPV2, mixtures of N-acyl prolines and N-acyl tyrosines were agonists and at TRPV4 N-acyl tryptophan and N-acyl tyrosine mixtures were agonists [39]. Table 3.2 provides a detailed list of which N-acyl amides act as ligands at TRPVs.

3.3.1 Activators of other TRP channels

Another subtype of TRP receptor is the ankyrin-type, known as the TRPA family. TRPA1, the "mustard-oil" receptor is expressed in sensory neurons, and topical application of TRPA1 agonists excites these nerve fibers to produce acute pain. However, desensitization of TRPA1 occurs rapidly after application of agonists and desensitization of TRPA1 is thought to underlie some of the antinociceptive effects of cannabinoids. In transfected HEK cells, the phytocannabinoids CBD and CBN were agonists at TRPA1 [24]. Interestingly, olive oil can activate TRPA1, which is responsible for some of its distinctive taste. Oleocanthal is hypothesized to be responsible for this phenomenon, as purified oleocanthal acted as an agonist in TRPA1 transfected cell lines [40]. Work by the Bradshaw group showed that *N*-acyl amides are also present at varying concentrations in olive oil, as well as in other cooking oils (unpublished data, but presented at ICRS 2013). It is not yet known if any of these endogenous *N*-acyl amides are also ligands at TRPA1 and if these lipids contribute to flavor in any perceivable way.

Yet another family of TRP receptors is present in the mammalian nervous system, called TRPC. One subtype of TRPC is TRPC5, which is known to be activated by lipids and is also known to be activated by nitric oxide, and plays a role in the neuronal signaling of pain. A putative ligand for TRPC5 is *N*-palmitoyl glycine (PalGly), which is present in high concentrations in the rat skin and spinal cord and is produced upon firing of sensory neurons. In the rat dorsal horn, P-Gly robustly inhibits heat-evoked firing of sensory neurons [41]. Given the similarities between the effects of PalGly and the effects of TRPC5 agonists, it is possible that PalGly activates the receptor TRPC5. Whether PalGly is an agonist at TRPC5 specifically remains to be tested [8].

3.3.2 Enzymes of the eCB System: A Common Link Between Cannabimimetic Endogenous Lipids

A potential link for the overlap of *N*-acyl amides, eCBs, and phytocannabiniod responses lies in their intersections at the level of biosynthetic and metabolic enzymes. There are three proposed separate pathways to yield AEA: the first is by direct hydrolysis of NAPE by a NAPE-specific phospholipase D (NAPE-PLD) [42]. In the second pathway, NAPE-specific phospholipases A_1/A_2 convert NAPE into lyso-NAPE and glycerophosphoAEA, which can then be hydrolyzed into AEA by a specific PLD [43]. The third pathway involves the generation of phospho-AEA from NAPE catalyzed by a NAPE-specific

PLC. A lipid phosphatase can then dephosphorylate to make AEA [44]. Fatty acid amide hydrolase (FAAH) is hypothesized to be responsible for the majority of AEA hydrolysis in the rodent brain. Even though FAAH is best known for its role in AEA hydrolysis, it can also hydrolyze other N-acyl amides [45]. N-Acyl amide structural analogs of AEA are hypothesized to be synthesized and broken down in similar pathways; however, there are likely additional modifications as outlined earlier in the case of NAGly and NADA. It may be that AEA is a precursor for a wide range of N-arachidonyl amides and its regulation would, therefore, have broad-reaching effects on the entire system of signaling lipids.

Demonstrating how the metabolism of eCBs that activate classical cannabinoid receptors is linked to the metabolism of noncannabinoid structural analogs, deleting the enzymes for eCB synthesis and breakdown can have wide-reaching effects on both categories of lipid signaling molecules. For example, studies with DAGL KO mice showed that functioning DAGL is necessary to maintain levels of arachidonic acid in the central nervous system [46]. When examining FAAH KO mice, it becomes apparent that the enzyme FAAH regulates much more than AEA. Although whole brain analysis revealed that the FAAH KO mice had 15 times as much AEA compared to WT, brain levels of OEA and PEA were also upregulated in the KO mice [47]. Using HPLC/MS/MS, Han et al. investigated the effects of acute, systemic injection of the FAAH inhibitor URB597 on the levels of several eicosanoid N-acyl amides in the mouse brain. As expected, there was a dose-dependent increase in levels of AEA. Interestingly, NAGly and A-GABA levels significantly decreased after URB597 treatment. These results suggest that the FAAH-mediated metabolism of AEA is necessary to maintain levels of other arachidonoyl acyl amides [48]. The link between eCBs and the wider eicosanoid signaling system is also demonstrated by the effects of MAGL blockade on levels of several arachidonic acid derivatives. After deleting MAGL in mice, Nomura et al. unsurprisingly reported a significant increase in 2-AG concentrations in whole-brain extracts. Levels of AEA remained constant. However, arachidonic acid levels significantly decreased in the brain of the KO animal. Interestingly, the MAGL KO mice demonstrated a significant attenuation of levels of certain eicosanoids, such as PGE_2, PGD_2, and $PGF_{2\alpha}$ [49].

Phytocannabinoids can indirectly influence eCBs by modifying the activity of eCB-system enzymes. For example, phytocannabinoids can regulate the activity of FAAH and, therefore, large families of cannabimimetic N-acyl amides. For example, CBD was able to inhibit radiolabeled AEA degradation into ethanolamine in neuronal membranes from mouse neuroblastoma cells with an IC_{50} of 27.5 μM [50]. De Petrocellis' group performed a similar experiment to assess the effects of phytocannabinoids on FAAH. However, they used rat brain membranes, which expressed FAAH as the only enzyme that hydrolyzed AEA. In their assays, CBD was the only compound able to inhibit FAAH [24]. Although FAAH is able to hydrolyze the other NAEs in a similar manner to AEA, it has recently been reported that the lysosomal enzyme N-acyl ethanolamine-hydrolyzing acid amidase (NAAA) breaks down NAEs, with a preference for PEA *in vitro*. NAAA shares no sequence homology with FAAH and, due to its preference for lysosomes, is only active at an acidic pH [51]. To assess the effects of phytocannabinoids as NAAA inhibitors, De Petrocellis et al. used HEK-293 cells expressing human NAAA and measured the conversion of radiolabeled PEA to radiolabeled ethanolamine. CBDA was the only phytocannabinoid that inhibited NAAA in its pure form, with an IC_{50} of 23 μM [24]. The effects of phytocannabinoids on MAGL have been investigated: CBC inhibited MAGL with an IC_{50} of 50.1 μM and CBG inhibited the enzyme with an IC_{50} of 95.7 μM [24]. Interestingly, an *in vitro* study showed that CBDA is a potent inhibitor of cyclooxygenase (COX) enzymes, with ninefold selectivity for COX-2. COX enzymes are required for the production of prostaglandins [52]. Oxidative metabolism of AEA is possible and involves several enzymes: cyclooxygenase-2 (COX-2), 12- and 15-lipoxygenases (LOX), and cytochrome P450 (CYP450). Although most oxygenated derivatives have no affinity for CB receptors, they can act as FAAH inhibitors to potentiate AEA's activity [30].

3.3.3 Effects of N-Acyl Amides on Additional Cannabinoid-Activated Proteins

3.3.3.1 Ca$_V$3

THC and CBD have both been shown to inhibit Ca$_V$3.1, Ca$_V$3.2, and Ca$_V$3.3 type calcium channels [53]. The Ca$_V$3 (also known as T-type) family of calcium channels is implicated in several clinical conditions, such as insomnia, epilepsy, neuropathic pain, and hypertension. Cazade et al. investigated whether there are endogenous lipids that inhibit T-type calcium channels. AEA was able to inhibit $Ca_V3.3$ currents in transfected HEK cell lines. Additionally, N-docosahexaenoyl glycine was a potent inhibitor of $Ca_V3.3$ currents; NAGly and N-linoleoyl glycine also inhibited the currents of the same channel, but were less potent. Other arachidonic acid derived N-acyl amides also inhibited $Ca_V3.3$ currents: N-arachidonoyl GABA, serine, alanine, serotonin, taurine, and dopamine all inhibited these currents. NADA was also able to inhibit $Ca_V3.1$ and 3.2 currents. NAGly was also able to increase the recovery time of $Ca_V3.3$ currents after inactivation [54].

3.3.3.2 VDAC1

Voltage-dependent anion channel 1 (VDAC1) is located on the outer membrane of mitochondria and is required for mitochondria-induced apoptosis. Inducing apoptosis is important for controlling the growth of cancer cells. The activity of CBD at VDAC1 was investigated, as CBD is known to induce apoptosis in breast carcinoma cells, BV2 microglial cells, and leukemic cell lines. Indeed, mass spectrometry analysis of fractionated BV2 cells confirmed that CBD accumulated in mitochondrial membranes and that CBD attenuated VDAC1 conductance [55]. Data from the Bradshaw laboratory indicated that five additional N-acyl amide families drive changes in calcium in BV-2 microglial cells, indicating that some of the activity at VDAC1 might be through this signaling mechanism [56].

3.4 N-ACYL AMIDE/eCBs WITH NOT YET IDENTIFIED PHYTOCANNABINOID LIGANDS

Finally, connections by activity associations are leading to discoveries of new signaling pathways for eCB structural analogs and the broader families of N-acyl amides. PEA and OEA both bind to the transcription factor peroxisome proliferator-activated receptor alpha (PPAR-α) to have analgesic, anti-inflammatory, and anorexic effects. In addition, N-stearoyl ethanolamine (SEA) also possesses anti-inflammatory and anorexic properties [57]. LEA can also activate PPAR-α in the intestine, where it may have a role in prolonging feeding latency [4]. OEA is an effective agonist of GPR119, a receptor highly expressed in the pancreas and colon [4]. The stimulation of GPR119 in the small intestine causes the release of glucagon-like peptide-1 (GLP-1), which enhances satiety and reduces food intake [58]. PPAR-α and GPR119 ligands are linked to the eCBs through activity profiles, shared precursor molecules, and structural similarities. Interestingly, these signaling pathways are continuously represented at the annual meeting on cannabinoid research, ICRS; however, there is no known phytocannabinoid link as yet. This is one of the many examples of how the interplay between the (Cannabis) plant and mammalian world continue to work together.

3.5 NOWHERE TO GO BUT UP

There is no doubt that the discover of Ananadamide sparked the field of research of *N*-acyl amides and their many signaling pathways and while it is equally likely that there are many of these molecules that are proving to be additional eCBs it is more likely that the understanding of the signaling roles of *N*-acyl amides as a family will surpass the confines of the cannabinoid signaling system. Like so many individuals that come from smaller beginnings and grow and branch out to encompass of wider community, there is also no doubt that those that continue to study this fascinating *N*-acyl amide group of molecules that we will always be grateful for such a rich environment from which to start.

REFERENCES

[1] Dumlao DS, Buczynski MW, Norris PC, Harkewicz R, Dennis EA. High-throughput lipidomic analysis of fatty acid derived eicosanoids and *N*-acylethanolamines. Biochim Biophys Acta 2011;1811:724–36.

[2] Astarita G, Geaga J, Ahmed F, Piomelli D. Targeted lipidomics as a tool to investigate endocannabinoid function. Int Rev Neurobiol 2009;85:35–55.

[3] Kuehl FA, Jacob TA, Ganley OH, Ormond RE, Meisinger MAP. The identification of *N*-(2-hydroxyethyl)-palmitamide as a naturally occurring anti-inflammatory agent. J Am Chem Soc 1957;79:5577–8.

[4] Ezzili C, Otrubova K, Boger DL. Fatty acid amide signaling molecules. Bioorg Med Chem Lett 2010;20:5959–68.

[5] Console-Bram L, Marcu J, Abood ME. Cannabinoid receptors: nomenclature and pharmacological principles. Prog Neuropsychopharmacol Biol Psychiatry 2012;38:4–15.

[6] Lin S, Khanolkar AD, Fan P, Goutopoulos A, Qin C, Papahadjis D, et al. Novel analogues of arachidonylethanolamide (anandamide): affinities for the CB1 and CB2 cannabinoid receptors and metabolic stability. J Med Chem 1998;41:5353–61.

[7] Tortoriello G, Rhodes BP, Takacs SM, Stuart JM, Basnet A, Raboune S, et al. Targeted lipidomics in *Drosophila melanogaster* identifies novel 2-monoacylglycerols and *N*-acyl amides. PLoS One 2013;8:e67865.

[8] Bradshaw HB, Raboune S, Hollis JL. Opportunistic activation of TRP receptors by endogenous lipids: exploiting lipidomics to understand TRP receptor cellular communication. Life Sci 2013;92(8–9):404–9.

[9] Pertwee RG, Thomas A, Stevenson LA, Ross RA, Varvel SA, Lichtman AH, et al. The psychoactive plant cannabinoid, Delta9-tetrahydrocannabinol, is antagonized by Delta8- and Delta9-tetrahydrocannabivarin in mice *in vivo*. Br J Pharmacol 2007;150:586–94.

[10] Cascio MG, Gauson LA, Stevenson LA, Ross RA, Pertwee RG. Evidence that the plant cannabinoid cannabigerol is a highly potent alpha2-adrenoceptor agonist and moderately potent 5HT1A receptor antagonist. Br J Pharmacol 2010;159:129–41.

[11] Howlett AC, Barth F, Bonner TI, Cabral G, Casellas P, Devane WA, et al. International Union of Pharmacology. XXVII. Classification of cannabinoid receptors. Pharmacol Rev 2002;54:161–202.

[12] Mchugh D, Hu SS, Rimmerman N, Juknat A, Vogel Z, Walker JM, et al. *N*-Arachidonoyl glycine, an abundant endogenous lipid, potently drives directed cellular migration through GPR18, the putative abnormal cannabidiol receptor. BMC Neurosci 2010;11:44.

[13] Pertwee RG. Receptors and channels targeted by synthetic cannabinoid receptor agonists and antagonists. Curr Med Chem 2010;17:1360–81.

[14] Pertwee RG, Howlett AC, Abood ME, Alexander SP, Di Marzo V, Elphick MR, et al. International Union of Basic and Clinical Pharmacology. LXXIX. Cannabinoid receptors and their ligands: beyond CB(1) and CB(2). Pharmacol Rev 2010;62:588–631.

[15] Jarai Z, Wagner JA, Varga K, Lake KD, Compton DR, Martin BR, et al. Cannabinoid-induced mesenteric vasodilation through an endothelial site distinct from CB1 or CB2 receptors. Proc Natl Acad Sci USA 1999;96:14136–41.

[16] Mchugh D, Page J, Dunn E, Bradshaw HB. Delta(9)-tetrahydrocannabinol and *N*-arachidonyl glycine are full agonists at GPR18 receptors and induce migration in human endometrial HEC-1B cells. Br J Pharmacol 2012;165:2414–24.

[17] Bradshaw HB, Rimmerman N, Hu SS, Benton VM, Stuart JM, Masuda K, et al. The endocannabinoid anandamide is a precursor for the signaling lipid *N*-arachidonoyl glycine by two distinct pathways. BMC Biochem 2009;10:14.

[18] Sheskin T, Hanus L, Slager J, Vogel Z, Mechoulam R. Structural requirements for binding of anandamide-type compounds to the brain cannabinoid receptor. J Med Chem 1997;40:659–67.

[19] Bradshaw HB, Rimmerman N, Hu SS, Burstein S, Walker JM. Novel endogenous *N*-acyl glycines identification and characterization. Vitam Horm 2009;81:191–205.

[20] Di Marzo V, De Petrocellis L, Fezza F, Ligresti A, Bisogno T. Anandamide receptors. Prostaglandins Leukot Essent Fatty Acids 2002;66:377–91.

[21] Ligresti A, Moriello AS, Starowicz K, Matias I, Pisanti S, De Petrocellis L, et al. Antitumor activity of plant cannabinoids with emphasis on the effect of cannabidiol on human breast carcinoma. J Pharmacol Exp Ther 2006;318:1375–87.

[22] De Petrocellis L, Vellani V, Schiano-Moriello A, Marini P, Magherini PC, Orlando P, et al. Plant-derived cannabinoids modulate the activity of transient receptor potential channels of ankyrin type-1 and melastatin type-8. J Pharmacol Exp Ther 2008;325:1007–15.

[23] Qin N, Neeper MP, Liu Y, Hutchinson TL, Lubin ML, Flores CM. TRPV2 is activated by cannabidiol and mediates CGRP release in cultured rat dorsal root ganglion neurons. J Neurosci 2008;28:6231–8.

[24] De Petrocellis L, Ligresti A, Moriello AS, Allara M, Bisogno T, Petrosino S, et al. Effects of cannabinoids and cannabinoid-enriched Cannabis extracts on TRP channels and endocannabinoid metabolic enzymes. Br J Pharmacol 2011;163:1479–94.

[25] De Petrocellis L, Orlando P, Moriello AS, Aviello G, Stott C, Izzo AA, et al. Cannabinoid actions at TRPV channels: effects on TRPV3 and TRPV4 and their potential relevance to gastrointestinal inflammation. Acta Physiol (Oxf) 2012;204:255–66.

[26] Ho KW, Ward NJ, Calkins DJ. TRPV1: a stress response protein in the central nervous system. Am J Neurodegener Dis 2012;1:1–14.

[27] Hwang SW, Cho H, Kwak J, Lee SY, Kang CJ, Jung J, et al. Direct activation of capsaicin receptors by products of lipoxygenases: endogenous capsaicin-like substances. Proc Natl Acad Sci USA 2000;97:6155–60.

[28] Craib SJ, Ellington HC, Pertwee RG, Ross RA. A possible role of lipoxygenase in the activation of vanilloid receptors by anandamide in the guinea-pig bronchus. Br J Pharmacol 2001;134:30–7.

[29] Di Marzo V, Bisogno T, De Petrocellis L. Anandamide: some like it hot. Trends Pharmacol Sci 2001;22:346–9.

[30] Kozak KR, Crews BC, Morrow JD, Wang LH, Ma YH, Weinander R, et al. Metabolism of the endocannabinoids, 2-arachidonylglycerol and anandamide, into prostaglandin, thromboxane, and prostacyclin glycerol esters and ethanolamides. J Biol Chem 2002;277:44877–85.

[31] Fowler CJ. Plant-derived, synthetic and endogenous cannabinoids as neuroprotective agents. Non-psychoactive cannabinoids, "entourage" compounds and inhibitors of N-acyl ethanolamine breakdown as therapeutic strategies to avoid pyschotropic effects. Brain Res Brain Res Rev 2003;41:26–43.

[32] Ross RA. Anandamide and vanilloid TRPV1 receptors. Br J Pharmacol 2003;140:790–801.

[33] Adamczyk P, Miszkiel J, Mccreary AC, Filip M, Papp M, Przegalinski E. The effects of cannabinoid CB1, CB2 and vanilloid TRPV1 receptor antagonists on cocaine addictive behavior in rats. Brain Res 2012;1444:45–54.

[34] Zygmunt PM, Petersson J, Andersson DA, Chuang H, Sorgard M, Di Marzo V, et al. Vanilloid receptors on sensory nerves mediate the vasodilator action of anandamide. Nature 1999;400:452–7.

[35] Huang SM, Bisogno T, Trevisani M, Al-Hayani A, De Petrocellis L, Fezza F, et al. An endogenous capsaicin-like substance with high potency at recombinant and native vanilloid VR1 receptors. Proc Natl Acad Sci USA 2002;99(12):8400–5.

[36] Hu SS, Bradshaw HB, Benton VM, Chen JS, Huang SM, Minassi A, et al. The biosynthesis of N-arachidonoyl dopamine (NADA), a putative endocannabinoid and endovanilloid, via conjugation of arachidonic acid with dopamine. Prostaglandins Leukot Essent Fatty Acids 2009;81:291–301.

[37] Chu CJ, Huang SM, De Petrocellis L, Bisogno T, Ewing SA, Miller JD, et al. N-Oleoyldopamine, a novel endogenous capsaicin-like lipid that produces hyperalgesia. J Biol Chem 2003;278:13633–9.

[38] Saghatelian A, Mckinney MK, Bandell M, Patapoutian A, Cravatt BF. A FAAH-regulated class of N-acyl taurines that activates TRP ion channels. Biochemistry 2006;45:9007–15.

[39] Frontiers in Cellular Neuroscience. http://www.frontiersin.org/Journal/Abstract. aspx?s=156&name=cellular_neuroscience&ART_DOI=10.3389/fncel.2014.00195; 2014.

[40] Peyrot Des Gachons C, Uchida K, Bryant B, Shima A, Sperry JB, Dankulich-Nagrudny L, et al. Unusual pungency from extra-Virgin olive oil is attributable to restricted spatial expression of the receptor of oleocanthal. J Neurosci 2011;31:999–1009.

[41] Rimmerman N, Bradshaw HB, Hughes HV, Chen JS, Hu SS, Mchugh D, et al. N-Palmitoyl glycine, a novel endogenous lipid that acts as a modulator of calcium influx and nitric oxide production in sensory neurons. Mol Pharmacol 2008;74:213–24.

[42] Di Marzo V, Fontana A, Cadas H, Schinelli S, Cimino G, Schwartz JC, et al. Formation and inactivation of endogenous cannabinoid anandamide in central neurons. Nature 1994;372: 686–91.

[43] Simon GM, Cravatt BF. Endocannabinoid biosynthesis proceeding through glycerophospho-N-acyl ethanolamine and a role for alpha/beta-hydrolase 4 in this pathway. J Biol Chem 2006;281:26465–72.

[44] Liu J, Wang L, Harvey-White J, Osei-Hyiaman D, Razdan R, Gong Q, et al. A biosynthetic pathway for anandamide. Proc Natl Acad Sci USA 2006;103:13345–50.

[45] Mulder AM, Cravatt BF. Endocannabinoid metabolism in the absence of fatty acid amide hydrolase (FAAH): discovery of phosphorylcholine derivatives of N-acyl ethanolamines. Biochemistry 2006;45:11267–77.

[46] Reisenberg M, Singh PK, Williams G, Doherty P. The diacylglycerol lipases: structure, regulation and roles in and beyond endocannabinoid signalling. Philos Trans R Soc Lond B Biol Sci 2012;367:3264–75.

[47] Cravatt BF, Demarest K, Patricelli MP, Bracey MH, Giang DK, Martin BR, et al. Supersensitivity to anandamide and enhanced endogenous cannabinoid signaling in mice lacking fatty acid amide hydrolase. Proc Natl Acad Sci USA 2001;98:9371–6.

[48] Han B, Wright R, Kirchhoff AM, Chester JA, Cooper BR, Davisson VJ, et al. Quantitative LC-MS/MS analysis of arachidonoyl amino acids in mouse brain with treatment of FAAH inhibitor. Anal Biochem 2013;432:74–81.

[49] Nomura DK, Morrison BE, Blankman JL, Long JZ, Kinsey SG, Marcondes MC, et al. Endocannabinoid hydrolysis generates brain prostaglandins that promote neuroinflammation. Science 2011;334:809–13.

[50] Bisogno T, Hanus L, De Petrocellis L, Tchilibon S, Ponde DE, Brandi I, et al. Molecular targets for cannabidiol and its synthetic analogues: effect on vanilloid VR1 receptors and on the cellular uptake and enzymatic hydrolysis of anandamide. Br J Pharmacol 2001;134:845–52.

[51] Ueda N, Tsuboi K, Uyama T. Metabolism of endocannabinoids and related N-acylethanolamines: canonical and alternative pathways. FEBS J 2013.

[52] Takeda S, Misawa K, Yamamoto I, Watanabe K. Cannabidiolic acid as a selective cyclooxygenase-2 inhibitory component in cannabis. Drug Metab Dispos 2008;36:1917–21.

[53] Ross HR, Napier I, Connor M. Inhibition of recombinant human T-type calcium channels by Delta9-tetrahydrocannabinol and cannabidiol. J Biol Chem 2008;283:16124–34.

[54] Cazade M, Nuss CE, Bidaud I, Renger JJ, Uebele VN, Lory P, et al. Cross-modulation and molecular interaction at the Cav3.3 protein between the endogenous lipids and the T-type calcium channel antagonist TTA-A2. Mol Pharmacol 2014;85:218–25.

[55] Rimmerman N, Ben-Hail D, Porat Z, Juknat A, Kozela E, Daniels MP, et al. Direct modulation of the outer mitochondrial membrane channel, voltage-dependent anion channel 1 (VDAC1) by cannabidiol: a novel mechanism for cannabinoid-induced cell death. Cell Death Dis 2013;4:e949.

[56] Raboune S, Stuart JM, Leishman E, Takacs SM, Rhodes B, Basnet A, et al. Novel endogenous N-acyl amides activate TRPV1-4 receptors, BV-2 microglia, and are regulated in brain in an acute model of inflammation. Front. Cell. Neurosci 2014;8:195.

[57] Astarita G, Piomelli D. Lipidomic analysis of endocannabinoid metabolism in biological samples. J Chromatogr B Analyt Technol Biomed Life Sci 2009;877:2755–67.

[58] Hansen KB, Rosenkilde MM, Knop FK, Wellner N, Diep TA, Rehfeld JF, et al. 2-Oleoyl glycerol is a GPR119 agonist and signals GLP-1 release in humans. J Clin Endocrinol Metab 2011;96:E1409–1417.

CHAPTER 4

Oxidative Metabolites of Endocannabinoids Formed by Cyclooxygenase-2

Lawrence J. Marnett, Philip J. Kingsley, Daniel J. Hermanson

The biochemistry and pharmacology of the two major endocannabi-noids (ECs), arachidonoylethanolamide (AEA) and 2-arachidonoylg-lycerol (2-AG), have been extensively investigated as described elsewhere in this volume [1,2]. They are generated on-demand from phospholipid precursors and are metabolically inactivated by hydrolysis to arachidon-ic acid (AA) (Figure 4.1). The balance between generation and hydro-lysis determines the extent of EC tone at cannabinoid (CB) and other receptors. The arachidonoyl group of both 2-AG and AEA contains four methylene-interrupted *cis* double bonds that are reactive to oxidiz-ing agents, either enzymatic or nonenzymatic. AA itself is a substrate for cytochromes P-450, lipoxygenases, and cyclooxygenases that introduce one atom, one molecule, and two molecules of O_2, respectively, into its carbon framework. The products of these oxygenations are either bio-active lipids or precursor to bioactive lipids. Included in this group are epoxyeicosatrienoic acids, dihydroxyeicosatetraenoic acids, leukotrienes,

The Endocannabinoidome: The World of Endocannabinoids and Related Mediators. DOI: 10.1016/B978-0-12-420126-2.00004-3
Copyright © 2015 Elsevier Inc. All rights reserved

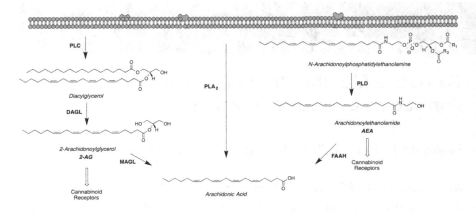

Fig. 4.1. Pathways of production of 2-AG, AA, and AEA.

lipoxins, protectins, prostaglandins, and thromboxanes. 2-AG and AEA are substrates for some of these fatty acid oxygenases and produce ul-timate metabolites that are analogous to the derivatives produced from AA except for the presence of the glycerol ester or ethanolamide moiety [3,4]. Oxidative metabolism of ECs has been much less studied than that of AA, but the available evidence suggests that under certain physiologi-cal or pathophysiological situations it can be an important source of lipid mediators. In addition, since the oxidative metabolites of ECs are not active at CB receptors, enzymatic oxidation may play an important role in regulating EC tone.

The field of oxidative metabolism of ECs is too large to review in the present chapter so we will focus on recent developments on their metab-olism by the fatty acid dioxygenase, cyclooxygenase-2 (COX-2). We will survey the analytical methods used for the detection and quantification of EC metabolites generated by COX-2, describe new tools with which to interrogate their formation and biological activities, then consider re-cent advances in our understanding of their biological roles.

4.1 COX-2 METABOLISM OF ENDOCANNABINOIDS

The two cyclooxygenase enzymes catalyze the *bis*-dioxygenation of AA to the prostaglandin endoperoxides, PGG_2 and PGH_2 [5,6]. PGH_2 dif-fuses from the COX protein and is converted by one or more of five endoperoxide-metabolizing enzymes to PGE_2, PGD_2, $PGF_{2\alpha}$, PGI_2, or

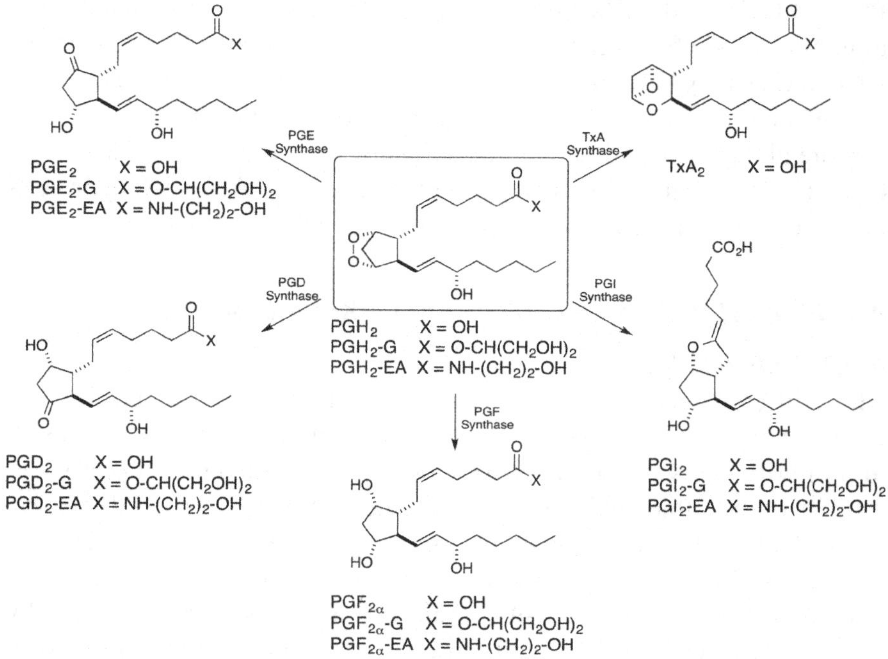

Fig. 4.2. Prostaglandin endoperoxide metabolism.

TxA$_2$ (Figure 4.2). PGH$_2$ metabolism occurs in a tissue-specific fashion depending on the balance of the various downstream synthases. TxA$_2$ and PGI$_2$ are unstable to spontaneous hydrolysis ($t_{1/2} \sim 30$ s) and produce the biologically inactive products, TxB$_2$ and 6-keto-PGF$_{1\alpha}$, respectively [7,8]. PGE$_2$, PGD$_2$, and PGF$_{2\alpha}$ are relatively stable chemically but they are rapidly metabolized to inactive derivatives by 15-keto-prostaglandin dehydrogenase [9]. Other metabolic steps such as β-oxidation and ω-oxidation occur more slowly. 2-AG and AEA are oxidized by COX-2 and, to a much lesser extent, by COX-1. They are oxidized to prosta-glandin endoperoxides containing glycerol ester or ethanolamide func-tionalities (i.e., PGG$_2$-G, PGH$_2$-G, PGG$_2$-EA, and PGH$_2$-AE) [10,11]. PGH$_2$-G and PGH$_2$-EA are metabolized by four of the five enzymes of endoperoxide metabolism (Figure 4.2). The exception is thromboxane synthase, which does not convert either compound to a TxA$_2$ analog ef-ficiently [12]. Thus, all of the attention on COX-2 metabolites of 2-AG and AEA has focused on the glycerol esters and ethanolamides of PGE$_2$,

PGD_2, $PGF_{2\alpha}$, and PGI_2. These prostaglandin glycerol esters and ethanol-amides are less effectively oxidized by 15-keto-prostaglandin dehydroge-nase so they may have the potential to circulate to tissues remote from the site of their generation without being metabolically inactivated [13]. This is more likely for the ethanolamide derivatives than the glycerol esters be-cause the latter can be hydrolyzed to prostaglandins. Although hydrolysis in blood is slow in humans, it appears to be rapid in rodents [13].

4.2 ANALYSIS OF ECS AND THEIR COX-2 METABOLITES

4.2.1 AEA and AG

Many different techniques have been reported for the analysis of AEA and 2-AG. HPLC with UV [14] or fluorescence [15] detection has been used, and in both cases, the analytes were derivatized with a chromophore or a fluorophore moiety. Yagen and Burstein employed a dansyl derivative of AEA [14] whereas Wang et al. used 4-(N,N-dimethylaminosulfonyl)-7-(N-chloroformylmethyl-N-methylamino)-2,1-3-benzoxadiazole   to derivatize both AEA and 2-AG [15]. The analytes were separated by reverse-phase (RP) HPLC columns. Gas chromatography-mass spec-trometry and gas chromatography-tandem mass spectrometry methods have also been reported [16,17]. Gas chromatographic methods require derivatization of the analytes.

Liquid chromatography-mass spectrometry (LC-MS) based meth-ods have rapidly evolved and represent the current state-of-the-art. Several methods have been reported based on selected ion monitoring (SIM) [18,19]. Typically, ECs are separated by RP chromatography us-ing isocratic or gradient elution and the mass spectrometer is operated in positive-ion atmospheric pressure chemical ionization (APCI) mode. Acetic acid is frequently added at low levels to promote the formation of the $[M + H]^+$ ion, which is used for quantification.

As triple quadrupole mass spectrometers have become a standard piece of laboratory equipment over the past decade, LC-MS/MS analysis of ECs has become commonplace. Indeed, multiple reports have been published describing the analytical methodology in significant detail [20–22]. In these methods, the ECs are subjected to RP chromatography similar to the systems described for LC-MS analyses. For example, Rich-ardson eluted AEA and 2-AG from an amide C8 column using a gradient containing 0.1% acetic acid to promote the formation of the [M + H]$^+$ ion. [M + H]$^+$ is typically the Q1 mass for both AEA (*m/z* 348) and 2-AG (*m/z* 379). *N*-Acylethanolamides produce a distinctive *m/z* 62 fragment on collisionally induced dissociation (CID) and this fragment was used as the Q3 mass by Richardson et al. and Zoerner et al. 2-AG generates several ions after CID but *m/z* 287 is most commonly used, being the Q3 mass employed by Zoerner and Thieme. As an alternative to protonation of the analyte, argentation or silver complexation, is performed to generate a [M + Ag]$^+$ cation [23]. This complex may be used as the Q1 mass in SRM and a loss of water (for AEA) or loss of the glycerol group (for 2-AG) can be used for the Q3 mass. In addition to the referenced methods, two reviews of EC analysis have been published [24,25].

4.2.2 Lipoaminoacids

AEA and 2-AG are the most studied amide or ester substrates for COX-2, but other endogenously occurring compounds have been identified [26,27]. These include the lipoaminoacids *N*-arachidonoyl glycine (NAGly), *N*-arachidonoyl alanine (NAla), and *N*-arachidonoyl-γ-aminobutryic acid (NAGABA) as well as the vanilloids *N*-arachidonoyldopamine (NADA), *O*-(3-methyl)-*N*-arachidonyldopamine (OMDA), and arvanil (Figure 4.1). *N*-Arachidonoyl amino acids and related conjugates have been studied almost exclusively with LC-MS techniques. The nonpolar arachidonoyl moiety of these compounds balances the polar amino acid residue so that the molecules are retained on RP columns. Lipoamino acids are amenable to ionization in both positive- and negative-ion modes. At neutral pH, the terminal carboxylate is deprotonated and a strong [M − H]$^-$ ion is generated. Positive-ion mode may be used in the analysis of *N*-arachidonoyl conjugates because at low pH (3–4) the terminal carboxylic acid is protonated, and the amide bond will complex with a proton to form a [M + H]$^+$. Finally, due to the high number of sites of unsaturation on the AA backbone, argentation is also an option.

Bradshaw et al. and Hu et al. describe negative-ion selected reaction monitoring (SRM) methods for the quantitation of NAGly and NADA [28,29]. In both cases the compounds were chromatographed via a gradient on a C18 column. The SRM transition for NAGly is *m/z* 360 → 74 and the transition for NADA is *m/z* 438 → 123. In each case, negative-ion mode was used and the Q1 mass corresponded to the [M − H]$^-$ ion. Han et al. described a positive-ion mode for the LC-MS/MS detection of NAGly, NAla, and NAGABA. Positive-ion SRM was employed and the analytes were chromatographed on a cyano-propyl column with a gradient of aqueous 10 mM ammonium acetate and methanol [30]. NAGly, NAla, and NAGABA are monitored via SRM with the [M + H]+ ion as the Q1 mass and m/z 287, 90, and 86 the Q3 masses, respectively.

4.2.3 Prostaglandin Glyceryl Ester (PG-G)

Several LC-MS methods have been reported for the analysis of the prostaglandin products of 2-AG oxygenation (Figure 4.1) by COX-2. In the initial reports of these compounds, Kozak et al. analyzed PGE_2-G, PGD_2-G, $PGF_{2\alpha}$-G, TXA_2-G, and TXB_2-G via LC-MS. The PG-Gs were separated by gradient elution from a C18 column and detected by SIM monitoring of the [M + Na]+ ions. Sodium complexation was promoted by including 0.001% sodium acetate in the mobile phase [10,12]. Kingsley et al. reported an LC-MS/MS method utilizing SRM for the analysis of PGE_2-G, PGD_2-G, $PGF_{2\alpha}$-G, and 6-keto-$PGF_{1\alpha}$-G [31]. Again, the analytes were eluted from a C18 column via a gradient and 2 mM ammonium acetate was added to the mobile phase to promote formation of $[M + NH_4]^+$ ions, which were the Q1 ions. The Q3 ions arose from the loss of a neutral ammonia and two water molecules for PGE_2-G, PGD_2-G, and $PGF_{2\alpha}$-G (m/z 391, 391, and 393, respectively) and the loss of a neutral ammonia and three water molecules for 6-keto-$PGF_{1\alpha}$-G (m/z 391). Hu et al. reported the presence of PGE_2-G in rat hindpaw using a similar LC-MS/MS method [32].

4.2.4 Prostaglandin Ethanolamides (PG-EAs or Prostamides)

Similar to PG-G analysis, almost all reports use LC-MS techniques with RP chromatography. Weber et al. reported the appearance of PGE_2-EA, PGD_2-EA, and $PGF_{2\alpha}$-EA in FAAH −/− mice by LC-MS/MS analysis of selected tissues. SRM in the positive-ion mode was used and the Q1 masses were the [M + H]+ ions. The Q3 masses were either ([M + H]+– $H_2O)^+$ or m/z 62 [33]. Koda et al. reported the analysis of PGD_2-EA in negative-ion mode. SIM of the [M-37]− ion was used for quantification [34]. Gatta et al. reported the presence of $PGF_{2\alpha}$-EA in rat spinal chord using mass spectrometry to identify $PGF_{2\alpha}$-EA [35]. The [M + Na]+ complex was used with SIM for quantitation and this complex underwent CID and time-of-flight analysis for species confirmation. Finally Ritter et al. quantified PGE_2-EA, PGD_2-EA, and $PGF_{2\alpha}$-EA via LC-MS/MS in positive-ion SRM mode. Again, the $[M + H]^+$ ion was the Q1 species and the Q3 was either $([M + H]^+-H_2O)^+$ or m/z 62 [36].

Little work has been done on the COX-2 oxygenated products of N-arachidonoyl conjugates. However, Prusakiewicz characterized the singly oxygenated HpETE products of NAGly and NAGABA in positive-ion mode by AG-coordination [27]. In a later report, Prusakiewicz used negative-ion mode to perform CID experiments on COX-2 metabolites of NAGly, NAla, and NAGABA.

Quantitation of ECs and their metabolites is commonly accomplished by stable isotope dilution. The stably labeled analog (or internal standard) of interest is added to the sample prior to purification and is monitored in the mass spectrometry method. The ratio of the analyte to its internal standard is used to calculate the amount of analyte present in the sample. Isotopically labeled internal standards are commercially available for many analytes, including AEA, 2-AG, and NAGly. Also, PG-G and PG-EA internal standards may be synthesized by relatively simple procedures [31]. Alternatively, some researchers construct a calibration curve for the analyte of interest. For example, Hu et al. quantitated PGE_2-G in rat hindpaw by comparison to a standard curve prepared with authentic material [32].

Many of the oxygenated products of COX-2 are present at vanishingly low levels in animals. In the future, one can expect highly sensitive tandem mass spectrometers to be employed in their analysis. For example, the combination of a SCIEX QTrap 6500 (perhaps utilizing MRM3 scanning) combined with a UPLC chromatographic regime may offer a limit of detection significantly lower than what has been reported, thereby offering researchers a method to further explore the biology of these compounds.

4.3 TOOLS TO STUDY PG-G AND PG-EA FUNCTIONS

Efforts to categorize the effects of EC-derived PG-EAs and PG-Gs are accelerating in part due to the availability of novel pharmacological tools. Direct analysis of $PGF_{2\alpha}$-EA functionality has been performed using a stable analog, bimatoprost (Figure 4.3). In contrast to $PGF_{2\alpha}$,

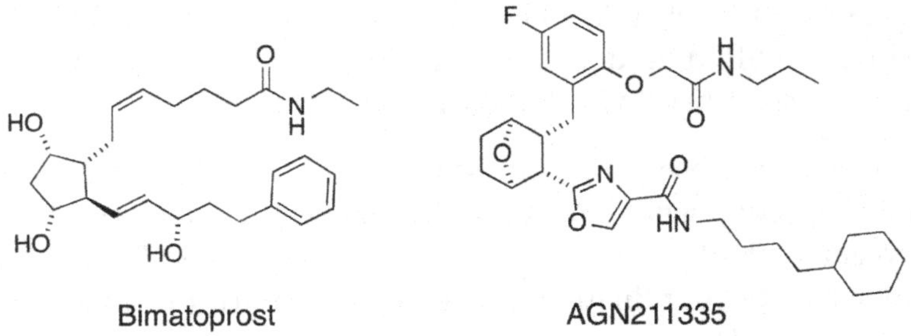

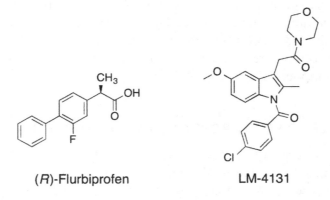

Fig. 4.3. Structures of bimatoprost and AGN211335.

which acts via the G protein-coupled FP receptor, $PGF_{2\alpha}$-EA, and bimatoprost act on a heterodimer consisting of a wild-type monomer and a splice variant of FP monomer [37–41]. AGN211335 is a selective antagonist of bimatoprost and $PGF_{2\alpha}$-EA action at FP/FP splice variant heterodimer receptors and it lacks antagonism of FP homodimers. Thus, it has been validated and utilized to study $PGF_{2\alpha}$-EA signaling [41].

While pharmacological probes for the FP/FP splice variant receptor heterodimer have been developed and validated, little is known about the receptors for other PG-EAs and PG-Gs. However, a series of studies have identified a class of "substrate-selective" COX-2 inhibitors that selectively block the oxygenation of 2-AG and AEA into PG-Gs and PG-EAs without inhibiting the oxidation of AA to PGs by COX-2 (Figure 4.4) [42–46]. These substrate-selective inhibitors are able to inhibit the production of PG-Gs and PG-EAs, leading to an increase

Fig. 4.4. Structures of substrate-selective COX-2 inhibitors.

in 2-AG and AEA, with no inhibition of PGs or modulation of AA levels [43,45]. Thus, the biological actions of PG-Gs and PG-EAs can be dissected from those of PGs using substrate-selective inhibitors compared to traditional nonsteroidal anti-inflammatory drugs (NSAIDs), which inhibit the production of PG-Gs, PG-EAs, and PGs by COX-1 and COX-2.

4.4 BIOLOGICAL EFFECTS OF PG-GS AND PG-EAS

Several studies have identified some of the effects of PG-Gs and PG-EAs but a comprehensive picture of their activities has not yet emerged. Importantly, PG-Gs and PG-EAs have biological functions that are distinct from traditional prostanoids and prostanoid receptors. PGE_2 and glycerol do not mobilize Ca^{2+} or activate PKC in RAW 264.7 cells, whereas PGE_2-G mobilizes Ca^{2+} and activates PKC in a concentration-dependent manner [47]. PGD_2-G and $PGF_{2\alpha}$-G do not modulate Ca^{2+} levels in the same cell line. $PGF_{2\alpha}$, which activates the FP receptor, also mobilizes Ca^{2+} from intracellular stores [48–50]. PGE_2-G exhibits no binding to the EP_2, DP, FP, IP, or TP receptors but binds to the EP_1, EP_3, EP_4 receptors with an affinity at least two orders of magnitude lower than that of PGE_2. Thus, PGE_2-G's mobilization of Ca^{2+} appears independent of binding to prostanoid receptors, including those known to mobilize intracellular Ca^{2+} (EP_1, EP_3, and FP). Further characterization established that PGE_2-G increases IP_3 levels, which can activate IP_3 receptors in the endoplasmic reticulum to release Ca^{2+} from intracellular stores; $PGF_{2\alpha}$ also increases IP_3 [51,52]. Additionally, PGE_2-G and PG-$F_{2\alpha}$ activate PKC, whereas PGE_2 does not. The increases in intracellular Ca^{2+} produced by PGE_2-G lead to an increase in activation of the ERK/ mitogen-activated protein kinase (MAPK) signaling pathway in a PLC, IP_3, and PKC-dependent manner.

Further characterization of the ability of prostaglandin analogs to elicit Ca^{2+} mobilization in RAW 264.7 cells revealed that glyceryl ester analogs of PGE_2 and $PGF_{2\alpha}$, including ester, thioester, and serinol amide linkages, cause a concentration-dependent increase in Ca^{2+} levels whereas prostaglandin ethanolamides do not [53]. This work also extended the ability of these compounds to mobilize calcium by identifying these effects in a human nonsmall cell lung cancer cell line (H1819).

These effects are independent of hydrolysis to PGE_2 or $PGF_{2\alpha}$, again suggesting that these molecules activate a set of receptors distinct from traditional prostanoid receptors.

While limited reports have detected PGE_2-G *in vivo*, one has identified PGE_2-G in rat footpad. Interestingly, PGE_2-G production was unchanged following treatment with carrageenan, an inflammatory stimulus, while PGE_2 levels were robustly increased [32]. Direct treatment of rat footpad with PGE_2-G or PGE_2 results in thermal hyperalgesia and mechanical allodynia. However, while the effects of PGE_2 are dependent on prostanoid receptors, PGE_2-G produces thermal hyperalgesia and mechanical allodynia when combined with a cocktail of prostanoid receptor antagonists. Additionally, both PGE_2 and PGE_2-G cause activation of NF-κB at low doses and suppression of NF-κB at higher doses. Again, these effects were dependent on different signaling pathways, as combination of PGE_2 with a cocktail of prostanoid receptor antagonists abrogated the modulation of NF-κB whereas PGE_2-G maintained its modulation of NF-κB in the presence of the prostanoid receptor antagonist cocktail.

A series of studies have identified PG-Gs and PG-EAs as important modulators of neurotransmission. PGE_2-G induces a concentration-dependent increase in the frequency of miniature inhibitory postsynaptic currents (mIPSCs) in GABAergic primary cultured hippocampal neurons [54]. This is in contrast to 2-AG, which decreases the frequency of mIPSCs [55]. Additional experiments identified that PGD_2-G, $PGF_{2\alpha}$-G, and PGD_2-EA increase the frequency of mIPSCs, while PGE_2-EA and $PGF_{2\alpha}$-EA do not. The increase in the frequency of mIPSCs is distinct from the effects of AA-derived prostanoids and is not mediated by CB_1 receptors [56].

PGE_2-G can also enhance miniature excitatory postsynaptic currents (mEPSCs) in glutamatergic neurons [57]. This increase in mEPSCs leads to neuronal injury and death as indicated by caspase 3 activation and terminal transferase-dUTP staining. These effects are mediated by ERK, p38 MAPK, IP_3, and NF-κB. Additionally, the PGE_2-G-induced neurotoxicity is mediated by NMDA receptors. PGE_2-G, PGE_2-EA, and $PGF_{2\alpha}$-EA elevate long-term potentiation in the hippocampus, whereas PGD_2-EA, PGD_2-G, and $PGF_{2\alpha}$-G do not [58]. The ERK, p38 MAPK, and IP3 pathways mediate the elevation of LTP elicited by PGE_2-G. Thus, EC-derived prostaglandins modulate synaptic signaling and synaptic plasticity in opposing manners to AEA and 2-AG and through distinct mechanisms from AA-derived prostaglandins.

A considerable amount of work has studied the pharmacology and biological effects of $PGF_{2\alpha}$-EA. As with PGE_2-G and PGE_2, $PGF_{2\alpha}$-EA and $PGF_{2\alpha}$ display distinct pharmacology. A primary motivation of this research has been to elucidate the mechanism of action of bimatoprost, which is clinically used for the treatment of glaucoma and eyelash hypotrichosis [59–61]. $PGF_{2\alpha}$ acts via the G protein-coupled FP receptor, whereas $PGF_{2\alpha}$-EA acts on a heterodimer consisting of wild type and a splice variant of FP [37–41]. AGN211335, a selective antagonist for the FP/FP splice variant heterodimer over FP homodimers has been useful in this setting [41].

A series of studies established that bimatoprost decreases periorbital fat deposits in glaucoma patients [62–65]. $PGF_{2\alpha}$ inhibits adipocyte differentiation through the activation of FP receptors, and subsequent activation of ERK 1 and 2, which results in phosphorylation and inactivation of PPARγ and calcium-calcineurin signaling pathways [66–68]. AEA stimulates adipogenesis through activation of the CB_1 receptor or PPARγ [69,70]. In contrast, $PGF_{2\alpha}$-EA and bimatoprost negatively regulate adipogenesis through their action at FP/FP splice variant heterodimer receptors, as identified by abrogation of their effects by AGN211335 [71]. Thus, $PGF_{2\alpha}$ and $PGF_{2\alpha}$-EA both reduce adipogenesis but do so through distinct receptors. These findings are supported by the observation that (R)-flurbiprofen accelerates adipogenesis [71].

In addition to the ability of $PGF_{2\alpha}$-EA to negatively regulate adipogenesis, it also mediates nociceptive responses in the spinal cord. The induction of knee inflammation with kaolin/λ-carrageenan in mice leads to the production of $PGF_{2\alpha}$-EA in the dorsal horn of the spinal cord [35]. Direct spinal application of $PGF_{2\alpha}$-EA leads to an increase in the firing of nociceptive neurons and a reduction in hot plate paw withdrawal latency in mice. Treatment with the FP splice variant receptor antagonist AGN211336 abrogates the effects of $PGF_{2\alpha}$-EA, while treatment with the FP receptor antagonist AL8810 does not. $PGF_{2\alpha}$ also increases nociceptive neuron firing and reduces the threshold for paw withdrawal latency and AL8810, but not AGN211336, blocks these effects. Intriguingly, mice treated with kaolin/λ-carrageenan have reduced nociceptive neuron firing and a reduction in paw withdrawal latency when treated with AGN211336, but not AL8810. Taken together, these studies indicate a central role for COX-2-mediated oxygenation of AEA and 2-AG and the resultant products in the activation of nociceptive neurons and subsequent pain transmission.

Preliminary studies have also suggested that PG-EAs may modulate kidney function. Infusion of AEA into the medulla of the kidney leads to increased urine volume and sodium and potassium excretion in a COX-2-dependent manner, suggesting a PG-EA-mediated function, but the species mediating these effects has not been identified [36]. Credence has been given to the hypothesis that PGE_2-EA mediates these effects, as direct PGE_2-EA application to the medulla leads to reduced mean arterial pressure and increased renal blood flow.

While most studies have suggested that PG-EAs and PG-Gs are proinflammatory and nociceptive mediators, a recent report has identified PGD_2-G as anti-inflammatory. Inhibition of ABHD6, which hydrolyzes 2-AG to AA and glycerol, leads to an increase in the production of PGD_2-G through the consecutive actions of COX-2 and prostaglandin D synthase in stimulated macrophages [72]. Increasing PGD_2-G levels through ABHD6 inhibition or direct application of PGD_2-G results in a reduction of proinflammatory cytokines in multiple stimulated macrophage cell lines and *in vivo*. The anti-inflammatory effects of PGD_2-G are not mediated by CB receptors, PPARγ, or PPARα, suggesting PGD_2-G produces these effects through a distinct signaling pathway. Thus, a series of studies have identified that PG-EAs and PG-Gs have discrete functions that appear to be mediated by receptors distinct from classical PG receptors.

4.5 PERSPECTIVE

The discovery that COX-2 oxygenates ECs opened the door to a potentially large new class of bioactive lipids that are glycerol ester and ethanolamide analogs of PGs. In addition, it raised the possibility that COX-2 could play a role as an additional modulator of EC levels and tone at the CB receptors. Although COX-2 is expressed basally in relatively few tissues, it is highly inducible by an extraordinarily broad range of stimuli suggesting the potential for EC depletion in many acute and chronic physiological and pathophysiological situations. However, despite the potential importance of COX-2 oxygenation of ECs, knowledge of its role in cellular systems and *in vivo* is limited. 2-AG is the most abundant and kinetically most significant COX-2 substrate but its PG-G metabolites are hydrolytically unstable and are converted to PGs. AEA levels are much lower than 2-AG levels so detection of PG-EA metabolites requires very high sensitivity analytical methods which have only recently been described [35]. The biological effects of PG-Gs and PG-EAs are tantalizing in their scope and potency but no receptors have been identified so investigators are still guessing at PG-G and PG-EA molecular pharmacology.

Progress has been made recently in the development of tools to help unravel the complexity of COX-2 metabolism of arachidonoyl-containing lipids. Substrate-selective inhibitors are available that selectively inhibit COX-2 oxygenation of ECs without inhibiting the oxygenation of AA [42,43,45]. They have been used to detect the involvement of PG-Gs and PG-EAs in inflammation, pain, and adipocyte differentiation [35,71,72]. One anticipates that similar studies will be conducted in many different systems. A receptor antagonist has been synthesized and validated that specifically antagonizes signaling by $PGF_{2\alpha}$-EA and may serve as a prototype for other antagonists of PG-G and PG-EA signaling [40]. This would help to dissociate responses induced by EC-derived PGs from that by AA-derived PGs and would be useful for the evaluation of candidate PG-G and PG-EA receptors.

One anticipates there will be significant advances in the near future in the identification of PG-G and PG-EA receptors and signal transduction pathways. It is also likely that the identity of the hydrolase(s) responsible for conversion of PG-Gs to PGs will be established and inhibitors will be developed. These would be very useful reagents for probing PG-G-dependent responses. Finally, it is worth noting that virtually all of the attention to date has been focused on COX-2 oxygenation of 2-AG and AEA. In fact there are many lipoamino acids in the central nervous system and in other tissues, so a parallel will be needed to investigate the importance of COX-2 oxygenation in their metabolism and biology.

REFERENCES

[1] Di Marzo V. The endocannabinoid system: its general strategy of action, tools for its pharmacological manipulation and potential therapeutic exploitation. Pharmacol Res 2009;60(2): 77–84.

[2] Astarita G, Piomelli D. Lipidomic analysis of endocannabinoid metabolism in biological samples. J Chromatogr B 2009;877(26):2755–67.

[3] Rouzer CA, Marnett LJ. Endocannabinoid oxygenation by cyclooxygenases, lipoxygenases, and cytochromes P450: cross-talk between the eicosanoid and endocannabinoid signaling pathways. Chem Rev 2011;111(10):5899–921.

[4] Rouzer CA, Marnett LJ. Non-redundant functions of cyclooxygenases: oxygenation of endocannabinoids. J Biol Chem 2008;283(13):8065–9.

[5] Hamberg M, Samuelsson B. Detection and isolation of an endoperoxide intermediate in prostaglandin biosynthesis. Proc Natl Acad Sci USA 1973;70:899–903.

[6] Nugteren DH, Hazelhof E. Isolation and properties of intermediates in prostaglandins biosynthesis. Biochim Biophys Acta 1973;326:448–61.

[7] Hamberg M, Svensson J, Samuelsson B. Thromboxanes: a new group of biologically active compounds derived from prostaglandin endoperoxides. Proc Natl Acad Sci USA 1975;72:2994–8.

[8] Whittaker N, Bunting S, Salmon J, et al. The chemical structure of prostaglandin X (prostacyclin). Prostaglandins 1976;12(6):915–28.

[9] Kung-Chao DT, Tai HH. NAD+-dependent 15-hydroxyprostaglandin dehydrogenase from porcine kidney. I. Purification and partial characterization. Biochim Biophys Acta 1980;614(1):1–13.

[10] Kozak KR, Rowlinson SW, Marnett LJ. Oxygenation of the endocannabinoid, 2-arachidonylglycerol, to glyceryl prostaglandins by cyclooxygenase-2. J Biol Chem 2000;275(43):33744–9.

[11] Yu M, Ives D, Ramesha CS. Synthesis of prostaglandin E_2 ethanolamide from anandamide by cyclooxygenase-2. J Biol Chem 1997;272:21181–6.

[12] Kozak KR, Crews BC, Morrow JD, et al. Metabolism of the endocannabinoids, 2-arachidonylglycerol and anandamide, into prostaglandin, thromboxane, and prostacyclin glycerol esters and ethanolamides. J Biol Chem 2002;277(47):44877–85.

[13] Kozak KR, Crews BC, Ray L, Tai HH, Morrow JD, Marnett LJ. Metabolism of prostaglandin glycerol esters and prostaglandin ethanolamides *in vitro* and *in vivo*. J Biol Chem 2001;276:36993–8. 0021-9258.

[14] Yagen B, Burstein S. Novel and sensitive method for the detection of anandamide by the use of its dansyl derivative. J Chromatogr 2000;740(1):93–9.

[15] Wang Y, Liu Y, Ito Y, et al. Simultaneous measurement of anandamide and 2-arachidonoylglycerol by polymyxin B-selective adsorption and subsequent high-performance liquid chromatography analysis: increase in endogenous cannabinoids in the sera of patients with endotoxic shock. Anal Biochem 2001;294(1):73–82.

[16] Schmid PC, Schwartz KD, Smith CN, Krebsbach RJ, Berdyshev EV, Schmid HH. A sensitive endocannabinoid assay. The simultaneous analysis of *N*-acylethanolamines and 2-monoacylglycerols. Chem Phys Lipids 2000;104(2):185–91.

[17] Zoerner AA, Gutzki FM, Suchy MT, et al. Targeted stable-isotope dilution GC-MS/MS analysis of the endocannabinoid anandamide and other fatty acid ethanol amides in human plasma. J Chromatogr B 2009;877(26):2909–23.

[18] Di Marzo V, Goparaju SK, Wang L, et al. Leptin-regulated endocannabinoids are involved in maintaining food intake. Nature 2001;410(6830):822–5.

[19] Patel S, Carrier EJ, Ho WS, et al. The postmortal accumulation of brain N-arachidonyleth-anolamine (anandamide) is dependent upon fatty acid amide hydrolase activity. J Lipid Res 2005;46(2):342–9.

[20] Richardson D, Ortori C, Chapman V, Kendall D, Barrett D. Quantitative profiling of en-docannabinoids and related compounds in rat brain using liquid chromatography–tandem electrospray ionization mass spectrometry. Anal Biochem 2007;216–26.

[21] Zoerner AA, Batkai S, Suchy MT, et al. Simultaneous UPLC-MS/MS quantification of the endocannabinoids 2-arachidonoyl glycerol (2AG), 1-arachidonoyl glycerol (1AG), and anan-damide in human plasma: minimization of matrix-effects, 2AG/1AG isomerization and degra-dation by toluene solvent extraction. J Chromatogr B 2012;883-884:161–71.

[22] Thieme U, Schelling G, Hauer D, et al. Quantification of anandamide and 2-arachidonoylg-lycerol plasma levels to examine potential influences of tetrahydrocannabinol application on the endocannabinoid system in humans. Drug Test Anal 2014;6(1-2):17–23.

[23] Kingsley PJ, Marnett LJ. Analysis of endocannabinoids by Ag+ coordination tandem mass spectrometry. Anal Biochem 2003;8–15.

[24] Kingsley PJ, Marnett LJ. Analysis of endocannabinoids, their congeners and COX-2 metabo-lites. J Chromatogr B 2009;877(26):2746–54.

[25] Zoerner AA, Gutzki FM, Batkai S, et al. Quantification of endocannabinoids in biological systems by chromatography and mass spectrometry: a comprehensive review from an analyti-cal and biological perspective. Biochim Biophys Acta 2011 1811;(11):706–23.

[26] Prusakiewicz JJ, Kingsley PJ, Kozak KR, Marnett LJ. Selective oxygenation of N-arachidonylglycine by cyclooxygenase-2. Biochem Biophys Res Commun 2002;296(3):612–7.

[27] Prusakiewicz JJ, Turman MV, Vila A, et al. Oxidative metabolism of lipoamino acids and vanilloids by lipoxygenases and cyclooxygenases. Arch Biochem Biophys 2007;464(2): 260–8.

[28] Bradshaw HB, Rimmerman N, Hu SS, et al. The endocannabinoid anandamide is a precur-sor for the signaling lipid N-arachidonoyl glycine by two distinct pathways. BMC Biochem 2009;10:14.

[29] Hu SS, Bradshaw HB, Benton VM, et al. The biosynthesis of N-arachidonoyl dopamine (NADA), a putative endocannabinoid and endovanilloid, via conjugation of arachidonic acid with dopamine. Prostaglandins leukot Essent Fatty Acids 2009;81(4):291–301.

[30] Han B, Wright R, Kirchhoff AM, et al. Quantitative LC-MS/MS analysis of arachidonoyl amino acids in mouse brain with treatment of FAAH inhibitor. Anal Biochem 2013;432(2): 74–81.

[31] Kingsley PJ, Rouzer CA, Saleh S, Marnett LJ. Simultaneous analysis of prostaglandin glyc-eryl esters and prostaglandins by electrospray tandem mass spectrometry. Anal Biochem 2005;343(2):203–11.

[32] Hu SS, Bradshaw HB, Chen JS, Tan B, Walker JM. Prostaglandin E2 glycerol ester, an en-dogenous COX-2 metabolite of 2-arachidonoylglycerol, induces hyperalgesia and modulates NFkappaB activity. Br J Pharmacol 2008;153(7):1538–49.

[33] Weber A, Ni J, Ling KH, et al. Formation of prostamides from anandamide in FAAH knockout mice analyzed by HPLC with tandem mass spectrometry. J Lipid Res 2004;45(4): 757–63.

[34] Koda N, Tsutsui Y, Niwa H, Ito S, Woodward DF, Watanabe K. Synthesis of prostaglandin F ethanolamide by prostaglandin F synthase and identification of Bimatoprost as a potent inhibitor of the enzyme: new enzyme assay method using LC/ESI/MS. Arch Biochem Biophys 2004;424(2):128–36.

[35] Gatta L, Piscitelli F, Giordano C, et al. Discovery of prostamide F2alpha and its role in inflam-matory pain and dorsal horn nociceptive neuron hyperexcitability. PloS One 2012;7(2):e31111.

[36] Ritter JK, Li C, Xia M, et al. Production and actions of the anandamide metabolite prosta-mide E2 in the renal medulla. J Pharmacol Exp Ther 2012;342(3):770–9.

[37] Matias I, Chen J, De Petrocellis L, et al. Prostaglandin ethanolamides (prostamides): *in vitro* pharmacology and metabolism. J Pharmacol Exp Ther 2004;309(2):745–57.

[38] Chen J, Senior J, Marshall K, et al. Studies using isolated uterine and other preparations show bimatoprost and prostanoid FP agonists have different activity profiles. Br J Pharmacol 2005;144(4):493–501.

[39] Woodward DF, Krauss AH, Chen J, et al. The pharmacology of bimatoprost (Lumigan). Surv Ophthalmol 2001;45(Suppl 4):S337–45.

[40] Woodward DF, Krauss AH, Wang JW, et al. Identification of an antagonist that selectively blocks the activity of prostamides (prostaglandin-ethanolamides) in the feline iris. Br J Phar-macol 2007;150(3):342–52.

[41] Liang Y, Woodward DF, Guzman VM, et al. Identification and pharmacological characteriza-tion of the prostaglandin FP receptor and FP receptor variant complexes. Br J Pharmacol 2008;154(5):1079–93.

[42] Prusakiewicz JJ, Duggan KC, Rouzer CA, Marnett LJ. Differential sensitivity and mechanism of inhibition of COX-2 oxygenation of arachidonic acid and 2-arachidonoylglycerol by ibu-profen and mefenamic acid. Biochemistry 2009;48(31):7353–5.

[43] Duggan KC, Hermanson DJ, Musee J, et al. (R)-Profens are substrate-selective inhibitors of endocannabinoid oxygenation by COX-2. Nat Chem Biol 2011;7(11):803–9.

[44] Windsor MA, Hermanson DJ, Kingsley PJ, et al. Substrate-selective inhibition of cyclooxygenase-2: development and evaluation of achiral profen probes. ACS Med Chem Lett 2012;3(9):759–63.

[45] Hermanson DJ, Hartley ND, Gamble-George J, et al. Substrate-selective COX-2 inhibition decreases anxiety via endocannabinoid activation. Nat Neurosci 2013;16(9):1291–8.

[46] Windsor MA, Valk PL, Xu S, Banerjee S, Marnett LJ. Exploring the molecular determinants of substrate-selective inhibition of cyclooxygenase-2 by lumiracoxib. Bioorg Med Chem Lett 2013;23(21):5860–4.

[47] Nirodi CS, Crews BC, Kozak KR, Morrow JD, Marnett LJ. The glyceryl ester of prostaglan-din E2 mobilizes calcium and activates signal transduction in RAW264.7 cells. Proc Natl Acad Sci USA 2004;101(7):1840–5.

[48] Woodward DF, Fairbairn CE, Goodrum DD, Krauss AH, Ralston TL, Williams LS. Ca^{2+} transients evoked by prostanoids in Swiss 3T3 cells suggest an FP-receptor mediated response. Adv Prostaglandin Thromboxane Leukot Res 1991;21A:367–70.

[49] Anthony TL, Fujino H, Pierce KL, Yool AJ, Regan JW. Differential regulation of Ca(2+)-dependent Cl_ currents by FP prostanoid receptor isoforms in Xenopus oocytes. Biochem Pharmacol 2002;63(10):1797–806.

[50] Davis JS, Weakland LL, Weiland DA, Farese RV, West LA. Prostaglandin F2 alpha stimulates phosphatidylinositol 4,5-bisphosphate hydrolysis and mobilizes intracellular Ca^{2+} in bovine luteal cells. Proc Natl Acad Sci USA 1987;84(11):3728–32.

[51] Dawson AP. Calcium signalling: how do IP3 receptors work? Curr Biol 1997;7(9):R544–7.

[52] Marchant JS, Taylor CW. Cooperative activation of IP3 receptors by sequential binding of IP3 and Ca^{2+} safeguards against spontaneous activity. Curr Biol 1997;7(7):510–8.

[53] Richie-Jannetta R, Nirodi CS, Crews BC, et al. Structural determinants for calcium mobiliza-tion by prostaglandin E2 and prostaglandin F2alpha glyceryl esters in RAW 264.7 cells and H1819 cells. Prostaglandins Other Lipid Mediators 2010;92(1–4):19–24.

[54] Sang N, Zhang J, Chen C. PGE2 glycerol ester, a COX-2 oxidative metabolite of 2-arachidonoyl glycerol, modulates inhibitory synaptic transmission in mouse hippocampal neurons. J Physiol 2006;572(Pt 3):735–45.

[55] Freund TF, Katona I, Piomelli D. Role of endogenous cannabinoids in synaptic signaling. Physiol Rev 2003;83(3):1017–66.

[56] Chen C, Magee JC, Bazan NG. Cyclooxygenase-2 regulates prostaglandin E2 signaling in hippocampal long-term synaptic plasticity. J Neurophysiol 2002;87(6):2851–7.

[57] Sang N, Zhang J, Chen C. COX-2 oxidative metabolite of endocannabinoid 2-AG enhances excitatory glutamatergic synaptic transmission and induces neurotoxicity. J Neurochem 2007;102(6):1966–77.

[58] Yang H, Zhang J, Andreasson K, Chen C. COX-2 oxidative metabolism of endocannabinoids augments hippocampal synaptic plasticity. Molec Cellular Neurosci 2008;37(4):682–95.

[59] Woodward DF, Krauss AH, Chen J, et al. Pharmacological characterization of a novel antiglaucoma agent, Bimatoprost (AGN 192024). J Pharmacol Exp Ther 2003;305(2):772–85.

[60] Woodward DF, Phelps RL, Krauss AH, et al. Bimatoprost: a novel antiglaucoma agent. Cardiovasc Drug Rev 2004;22(2):103–20.

[61] Woodward JA, Haggerty CJ, Stinnett SS, Williams ZY. Bimatoprost 0.03% gel for cosmetic eyelash growth and enhancement. J Cosmetic Dermatol 2010;9(2):96–102.

[62] Aydin S, Isikligil I, Teksen YA, Kir E. Recovery of orbital fat pad prolapsus and deepening of the lid sulcus from topical bimatoprost therapy: 2 case reports and review of the literature. Cutan Ocul Toxicol 2010;29(3):212–6.

[63] Peplinski LS, Albiani Smith K. Deepening of lid sulcus from topical bimatoprost therapy. Optometry Vision Sci 2004;81(8):574–7.

[64] Yam JC, Yuen NS, Chan CW. Bilateral deepening of upper lid sulcus from topical bimatoprost therapy. J Ocular Pharmacol Ther 2009;25(5):471–2.

[65] Filippopoulos T, Paula JS, Torun N, Hatton MP, Pasquale LR, Grosskreutz CL. Periorbital changes associated with topical bimatoprost. Ophthal Plast Reconstr Surg 2008;24(4):302–7.

[66] Miller CW, Casimir DA, Ntambi JM. The mechanism of inhibition of 3T3-L1 preadipocyte differentiation by prostaglandin F2alpha. Endocrinology 1996;137(12):5641–50.

[67] Reginato MJ, Krakow SL, Bailey ST, Lazar MA. Prostaglandins promote and block adipogenesis through opposing effects on peroxisome proliferator-activated receptor gamma. J Biol Chem 1998;273(4):1855–8.

[68] Liu L, Clipstone NA. Prostaglandin F2alpha inhibits adipocyte differentiation via a G alpha q-calcium-calcineurin-dependent signaling pathway. J Cell Biochem 2007;100(1):161–73.

[69] Bouaboula M, Hilairet S, Marchand J, Fajas L, Le Fur G, Casellas P. Anandamide induced PPARgamma transcriptional activation and 3T3-L1 preadipocyte differentiation. Eur J Pharmacol 2005;517(3):174–81.

[70] Karaliota S, Siafaka-Kapadai A, Gontinou C, Psarra K, Mavri-Vavayanni M. Anandamide increases the differentiation of rat adipocytes and causes PPARgamma and CB1 receptor upregulation. Obesity 2009;17(10):1830–8.

[71] Silvestri C, Martella A, Poloso NJ, et al. Anandamide-derived prostamide F2alpha negatively regulates adipogenesis. J Biol Chem 2013;288(32):23307–21.

[72] Alhouayek M, Masquelier J, Cani PD, Lambert DM, Muccioli GG. Implication of the anti-inflammatory bioactive lipid prostaglandin D2-glycerol ester in the control of macrophage activation and inflammation by ABHD6. Proc Natl Acad Sci USA 2013;110(43):17558–63.

CHAPTER 5

N-Acyldopamines and N-Acylserotonins: From Synthetic Pharmacological Tools to Endogenous Multitarget Mediators

Luciano De Petrocellis, Vincenzo Di Marzo

5.1 INTRODUCTION

Following the discovery of *N*-acyl-amino acids (also known as "lipoaminoacids," see Chapter 3 of this book), and in particular of the *N*-acylglycines [1], it appeared possible that amides between fatty acids and other bioactive amines could be present in mammalian tissues. Yet, some of these putative endogenous lipids had already been synthesized earlier on, and their activity studied both *in vitro* and *in vivo*. This is typically the case for the *N*-acyldopamines (NADs) [2,3] and *N*-acylserotonins (NASs) [4,5], which had been first synthesized

The Endocannabinoidome: The World of Endocannabinoids and Related Mediators. DOI: 10.1016/B978-0-12-420126-2.00005-5
Copyright © 2015 Elsevier Inc. All rights reserved

and examined for some of their *in vitro* actions in 2000 [2,3] and 1998 [4,5], and identified for the first time in the brain in 2002 [6,7], and then in the intestine in 2011 [8]. Given the nature of these long chain fatty acid amides, the initially only synthetic compounds were investigated as potential inhibitors of the inactivation of the previously discovered endocannabinoid *N*-arachidonoyl-ethanolamine (anandamide) and its bioactive congeners, such as *N*-palmitoyl- and *N*-oleoyl-ethanolamine [10], as well as of the sleep-inducing factor, oleamide [9], by "fatty acid amide hydrolase" (FAAH) [9]. However, and especially after their identification in mammalian tissues, many other pharmacological properties have been revealed, both *in vitro* and *in vivo*, for these two families of "bioactive amides of fatty acids" (BAFAs) [5,9], particularly for their unsaturated and polyunsaturated members. This chapter describes the state of the art of our current knowledge of NADs and NASs, their potential or established biosynthetic and inactivation mechanisms, and their several molecular targets in mammals.

5.2 *N*-ACYLDOPAMINES

5.2.1 Synthesis, Discovery as Endogenous Mediators and Metabolic Pathways

A series of NADs was first synthesized in 2000 by Bisogno and colleagues [2], and a few more members of this family of lipids were prepared a year later by Bezuglov et al. [3]. The compounds were tested for their affinity for cannabinoid CB1 and CB2 receptors and their capability of inhibiting anandamide hydrolysis by a membrane preparation containing FAAH, or anandamide reuptake by intact RBL-2H3 cells [2]. The compounds inhibited anandamide hydrolysis by neuroblastoma N18TG2 cell membranes with low potency ($19 < IC_{50} < 100 \ \mu M$) and in a competitive manner, but appeared to exhibit higher activity at displacing the tritiated selective CB1 receptor ligand SR141716A from rat brain membranes ($250 < K_i < 3900$ nM), with the highest affinity ligand being *N*-arachidonoyldopamine (NADA) [2]. Importantly, NADs did not inhibit the binding of the tritiated CB1/CB2 receptor ligand, WIN55,212-2 to rat spleen membranes ($K_i > 10 \ \mu M$), which express high amounts of CB2 receptors. NADA did not displace high-affinity D1 and D2 dopamine-receptor ligands from rat brain membranes at concentrations up to 10 μM, thus suggesting that this compound has little affinity for these receptors [2]. NADA, and also the eicosapentaenoyl (20:5 omega 3), docosapentaenoyl (22:5 omega 3), α-linolenoyl (18:3 omega 3), and pinolenoyl (5c,9c,12c 18:3 omega 6) homologs were found to inhibit anandamide reuptake by RBL-2H3 basophilic leukemia and C6 glioma cells with intermediate potency (17.5 < IC_{50} < 33 μM). *N*-Docosapentaenoyl-dopamine exhibited fourfold selectivity for the anandamide transporter over FAAH [2]. Of all the NADs tested, NADA was the most potent and efficacious as a CB1 agonist, as assessed by measuring its stimulatory effect on intracellular Ca^{2+} mobilization in undifferentiated N18TG2 neuroblastoma cells, or its capability of inhibiting the proliferation of human breast MCF-7 cancer cells, two effects counteracted by a specific CB1 antagonist. Accordingly, NADA behaved as a CB1 agonist also *in vivo* by inducing hypothermia, hypolocomotion, catalepsy, and analgesia in mice (1–10 mg/kg) [2].

After this initial report, the pharmacology of NADs was not further investigated until their identification in the bovine and rat brain by Huang et al. and Chu et al., in 2002–2003 [6,7]. NADA exhibited the highest concentrations in the striatum, hippocampus, and cerebellum and the lowest in dorsal root ganglia, thus reflecting in some ways the concentrations of dopamine in these neural tissues. However, its amounts were never higher than 6.5 pmol/g, and hence approximately one order of magnitude lower than anandamide brain levels [6]. Other long chain NADs, such as *N*-oleoyl-, *N*-palmitoyl-, and *N*-stearoyl-dopamine (abbreviated as OLDA, PALDA, and STEARDA, respectively) appeared to be more abundant [7]. Importantly, NADs were found to either directly activate the transient receptor potential of vanilloid type-1 (TRPV1) cation channels (in the case of NADA and OLDA) [6,7], thus behaving as "endovanilloids," or to potentiate the effects at these channels of other endogenous compounds including NADA and anandamide (in the case of PALDA and STEARDA) [11]. Therefore, like anandamide, NADA is both an "endocannabinoid" (endogenous agonist of cannabinoid receptors) and an "endovanilloid" (endogenous activator of TRPV1 channels), although with relative functional potency (TRPV1 > CB1 > CB2) different from that of anandamide (CB1 > CB2 = TRPV1), whereas OLDA is selective for TRPV1 channels. On the other hand, PALDA and STEARDA act as "entourage compounds" for NADA and anandamide, similar to the previously shown N-palmitoylethanolamide/anandamide combination at TRPV1 and CB1 receptors [12,13], and the 2-palmitoyl- and 2-linoleoylglycerol/2-arachidonoylglycerol (2-AG) combinations at CB1 and CB2 receptors [14].

Later work also suggested the possible biosynthetic and catabolic routes for NADs. The initial hypothesis that NADA relies on endogenous dopamine for its formation, and might, therefore, be produced via direct condensation between dopamine and arachidonic acid [6], was confirmed in a later study. It was observed that (1) NADA synthesis requires tyrosine β-hydroxylase, the enzyme that converts tyrosine into dopamine in dopaminergic terminals; (2) N-arachidonoyltyrosine, which is also present in the brain as an endogenous lipid, is not an intermediate in NADA biosynthesis; (3) FAAH is involved in NADA biosynthesis from arachidonic acid and dopamine either as a rate-limiting enzyme that liberates arachidonic acid from anandamide, or as a conjugation enzyme, or both [15]. Also, recent studies carried out on OLDA [16] confirmed the preliminary observation that NADA, following its reuptake by cells mediated by the same mechanism allowing for anandamide reuptake [6], is not inactivated by FAAH-catalyzed hydrolysis, but rather by its conversion by catechol-O-methyl-transferase (COMT) to less active O-methyl-NADAs [6]. The authors showed that OLDA was methylated by COMT in all conditions studied, yielding the O-methylated derivatives. The methylation was reversed by tolcapone, a potent COMT inhibitor, in a dose-dependent manner [16]. Although the authors of this study suggested that methylation of OLDA may enhance its bioactive properties, including ability to interact with TRPV1 receptors, both 3- and, particularly, 4-O-methyl-NADA had been previously shown to be several-fold less potent than NADA at activating such channels [17]. Furthermore, in support of the role of COMT in the inactivation of endovanilloids such as NADA and OLDA, Cristino et al. showed that this enzyme and TRPV1 are strongly coexpressed in neurons of the hippocampus and cerebellar cortex [18]. Sulfation has also been suggested as a potential catabolic route for NADs, although such reaction requires high concentrations of substrates to occur *in vitro* [19]. Rat liver cytochrome P450 enzymes are capable of metabolizing the arachidonoyl chain of NADA to 19- and 20-hydroxy derivatives, which are less potent than the parent compound at activating TRPV1 [20]. Finally, rat 15-lipoxygenase-1, but not cyclooxygenase-2 (COX-2), can recognize NADA as substrate [21].

5.2.2 Molecular Targets
5.2.2.1 TRPV1- and/or CB1-Mediated Actions

NADA and OLDA have been shown to produce several biological effects, both *in vitro* and *in vivo*, via activation of TRPV1 channels. NADA potently activates native vanilloid receptors in neurons from rat dorsal root ganglia and hippocampus, thereby inducing the release of substance P and calcitonin gene-related peptide (CGRP) from dorsal spinal cord slices, and enhancing hippocampal paired-pulse depression, respectively [6]. As expected from TRPV1 agonists, intradermal NADA and OLDA also induce TRPV1-mediated thermal hyperalgesia [7]. NADA exerts TRPV1-mediated contractile responses in both the guinea pig bronchus and urinary bladder, to an extent dependent on pharmacodynamics and bioavailability [22]. NADA is a potent vasorelaxant of the rat mesenteric artery, and its mechanism of action depends on the segment of the artery used to investigate its effects. In the small mesenteric artery, the effects of NADA are mediated by stimulation of TRPV1 and an as yet uncharacterized endothelial CB receptor (which could be the "orphan" receptor GPR18, see below), whereas in the superior mesenteric artery, vasorelaxation is mediated through TRPV1 and CB1 receptors [23]. Also when investigating its effects on primary afferent fiber and spinal cord neuronal responses in the rat, NADA was found to produce excitatory responses via both CB1 and TRPV1 receptors, resulting in analgesic effects [24]. In dopaminergic neurons of the substantia nigra pars compacta, NADA was shown to either increase or reduce glutamatergic transmission onto dopaminergic neurons by activating either TRPV1 or CB1 receptors at low and high concentrations, respectively [25]. Instead, under conditions causing glutamate spillover in synapses onto the somas of dopaminergic neurons in the same area, NADA is endogenously produced and retrogradely inhibits evoked inhibitory postsynaptic currents in these somas uniquely via CB1 receptors [26]. Disinhibition of dopaminergic neurons in this area may have profound effects on locomotion, especially in neuromotor disorders accompanied by glutamate excitotoxicity.

Activation of TRPV1 by NADA in F11 neurons (which are similar to dorsal root ganglion neurons) was found to result in growth cone retraction and collapse and formation of varicosities along neuritis [27]. These changes were due to TRPV1-activation-mediated disassembly of microtubules and were partly Ca^{2+}-independent. Prolonged activation with very low doses (1 nM) of NADA resulted in the shortening of neurites in the majority of isolectin B4-positive dorsal root ganglia neurons. The authors postulated that TRPV1 activation plays an inhibitory role in sensory neuronal extension and motility by regulating the disassembly of microtubules, which might have a role in chronic pain [27].

In vivo, NADA, like anandamide, was shown to inhibit emesis in the ferret probably via activation of CB1 receptors and activation and desensitization of TRPV1 channels [28]. In rats, high sodium intake, probably via upregulation of mesenteric TRPV1 expression, was found to increase the sensitivity of blood pressure responses to NADA, the enhanced depressor effect of which was prevented by blockade of TRPV1 or CGRP, but not CB1, receptors [29]. Interestingly, upregulation of TRPV1 expression was recently shown to be accompanied by higher sensitivity to NADA and OLDA also in peripheral blood mononuclear cells of end-stage kidney disease patients. The two compounds, through this channel, then contribute to the death of these cells [30]. Anandamide and NADA initiate TRPV1-dependent delayed cell death also in neuron-like cells *in vitro*, via an apoptosis-like process independent of caspase activity [31].

The effects of NADA on thermal hyperalgesia were recently evaluated in rats with a unilateral hind paw carrageenan-induced inflammation [32]. Intrathecal injection of NADA (1.5–50 μg) caused dose-dependent antihyperalgesia, which was inhibited by the TRPV1 antagonist AMG9810, but not by CB1 antagonist/inverse agonist AM251, when a low concentration of NADA was used, and by both drugs when NADA was injected at the highest dose [32].

The agonist activity of OLDA at TRPV1 channels has also been the subject of some *in vitro* and *in vivo* studies. This compound was shown to activate and desensitize TRPV1 in rat trigeminal neurons and to evoked paw lifting/licking, which was significantly less sustained in TRPV1 knockout mice [33]. Very recently, the involvement of TRPV1 in the modulation of learning and memory processes by stress was evaluated in mice by assessing the effects of OLDA on long-term potentiation (LTP) of the lateral nucleus of the amygdala (LA) induced by high-frequency stimulation of external capsule fibers. LA-LTP was reduced in OLDA-treated slices derived from adult C57BL/6 control mice, but not in slices treated with the specific TRPV1 receptor antagonist AMG9810, or prepared from TRPV1 "knockout" mice. On the other hand, a short period of acute stress, i.e., exposure to a forced swim test significantly impaired LA-LTP; and OLDA rescued LA-LTP in control but not TRPV1-deficient mice [34]. Thus, it is possible that activation of TRPV1 underlies stress-induced attenuation of LA-LTP, and that OLDA protects the amygdala from this effect by desensitizing TRPV1, thereby producing alleviation from the negative cognitive effects of stress. Possibly relevant to this finding is the recent observation that plasma levels of OLDA are negatively correlated with post-traumatic stress disorder (PTSD) scores in trauma-exposed individuals with and without PTSD [35]. Finally, still in the brain, OLDA (0.01–1 µg/rat) was very recently shown to accelerate the incidence of seizures in pentylenetetrazole- and amygdala-induced kindling in male rats. Seizures were instead reduced by the TRPV1 antagonist AMG-9810, which counteracted the proconvulsant effect of OLDA amygdala-induced kindling [36].

5.2.2.2 Non-TRPV1-, non-CB1-Mediated Actions

Among the non-TRPV1 and non-CB1 mediated actions of NADs so far reported, the capability of inhibiting T-type calcium channels is the one that can be exerted at the lowest (submicromolar) concentrations, and was confirmed in at least two independent studies, especially for the unsaturated members of this family such as NADA [37,38]. It would be interesting to assess if this effect of NADs, together with TRPV1 activation/desensitization, underlies, to some extent, their antihyperalgesic actions (see above); or if it is at the basis of the observation that NADA, like anandamide, inhibits K^+-evoked Ca^{2+} entry and transmitter release in hippocampal neurons, an effect shown to occur independently of CB1 and TRPV1 receptors and possibly through direct Ca^{2+} channel blockade [39]. Indeed, inhibition of T-type Ca^{2+} channels is another effect that NADs share with anandamide and other endogenous BAFAs, and so is the capability of NADA to antagonize transient receptor potential of melastatin type-8 (TRPM8) channels [40].

NADA also potently inhibits (IC_{50} = 150 nM) the oxidation of arachidonic acid by platelet 12-lipoxygenase [21], thus potentially reducing the formation of 12-lipoxygenase-derived eicosanoids. Indeed, NADA, OLDA, and, particularly, PALDA inhibit the formation of inflammatory mediators in IgE-challenged RBL-2H3 basophil-like cells [41]. The authors suggested that such action could be due in part to *inhibition* of COX-2. On the other hand, in brain endothelial cells, NADA was found to activate a redox-sensitive p38 MAPK pathway that *stabilizes* COX-2 mRNA, resulting in the accumulation of the COX-2 protein [42]. This effect seems to be dependent on the dopamine moiety of the molecule and independent of CB1 and TRPV1 activation. NADA inhibited the expression of microsomal prostaglandin E synthase-1 and the release of PGE_2, but upregulated the expression of lipocalin-type prostaglandin D synthase, thereby enhancing PGD_2 release. It was suggested that the anti-inflammatory activity of NADA and other molecules of the same family might allow them to prevent blood–brain barrier injury under inflammatory conditions, thus providing potential to design novel therapeutic strategies for neuroinflammatory diseases [42]. Indeed, the same authors had previously demonstrated that (1) anandamide and NADA have opposite effects on glial cells, with the latter compound being capable of exerting potential antioxidative and anti-inflammatory actions through reduction of PGE_2 biosynthesis in lipopolysaccharide-activated microglia, without modifying the expression or enzymatic activity of COX-2 and the production of PGD_2 [43]; and (2) NADA and, less potently, OLDA inhibit activation of necrosis factor-kappa B, nuclear factor of activated T-cells, and activator protein-1 signaling pathways in human T cells, possibly by targeting an active form of the phosphatase calcineurin [44]. Of note, NADA is also a potent inhibitor of *N*-formyl-l-methionyl-l-leucyl-l-phenylalanine-induced migration of human neutrophils (IC_{50} = 8.8 nM) [45], and of proinflammatory mediator (nitric oxide, interleukins-1β and -6, and TNF-α) release from lipopolysaccharide-activated RAW264.7 macrophages (IC_{50} = 2 μM) [46]. The mechanisms of these effects were, however, not investigated.

Among other less-investigated non-TRPV1, non-CB1-mediated *in vitro* effects of NADA, the following can be listed: (1) vasorelaxation of the rat aorta via activation of peroxisome proliferator-activated receptor-γ (as well as CB1) [47]; (2) selective induction of oxidative stress-mediated cell death in hepatic stellate cells but not in hepatocytes [48]; (3) inhibition of the hydrolysis of the endocannabinoid 2-AG by monoacylglycerol lipase [49]; and (4) uncoupling of sarcoplasmic reticulum Ca^{2+}-ATPase [50]. On the other hand, OLDA was found to act as a partial agonist at the orphan G-protein-coupled receptor, GPR119, which plays an important role in the release of glucagon-like peptide-1 (GLP-1) from the small intestine [51].

5.2.3 N-Acyldopamine Signaling: Conclusions

In conclusion, NADs, and NADA and OLDA in particular, might be regarded as endogenous signals characterized by multiple extra- and intracellular mechanisms of action and "promiscuity" of molecular targets, which include both enzymes and often different types (GPCRs, PPARs, channels) of receptors. This is typical of many lipid mediators. However, it remains to be established if the low nanomolar tissue concentrations of NADs are compatible with their physiological action at receptors and proteins with which they have been shown to interact *in vitro* often only at high-nanomolar–low-micromolar concentrations. Future studies should address the possibility that the tissue levels of NADs are strongly elevated during pathological conditions, thus creating the possibility for their modulatory actions to occur also via targets other than CB1 and TRPV1.

5.3 N-ACYLSEROTONINS

5.3.1 Synthesis as FAAH Inhibitors and Recent Discovery as Endogenous Mediators

N-Arachidonoyl-serotonin (AA-5-HT) was first synthesized in 1998 as a potential FAAH inhibitor [4,5]. Indeed, the compound was found to behave as a noncompetitive and mixed inhibitor of anandamide hydrolysis by cell membranes expressing FAAH as well as by intact cells, at concentrations in the low-micromolar range, depending on the assay conditions [4,52]. After its development, and despite it being not such a potent FAAH inhibitor, AA-5-HT was shown to elevate both anandamide and 2-AG levels in the brain, when administered subchronically to rats [53]. AA-5-HT was also shown to be resistant to hydrolysis to serotonin *in vivo* [53], and to be devoid, like other FAAH inhibitors, of strong activity in the "tetrad" of behavioral actions, which is instead indicative of "direct" CB1 receptor agonism [4]. Therefore, the synthetic compound was widely used thereafter as a pharmacological tool in several investigations on the role of the enzyme in the control of endocannabinoid levels under both physiological and pathological conditions. It was shown to elevate anandamide and/or 2-AG levels in many cell types and tissues [54–63]. Although structure activity studies were performed in order to improve its potency by modifying its chemical structure [53,64], AA-5-HT remains the most potent FAAH inhibitors in its family of lipids.

A key, as well as serendipitous, discovery was the finding in 2007, that AA-5-HT is also a potent (IC_{50} in the 40 nM range of concentrations) TRPV1 antagonist [65]. This led to the reevaluation of this compound, not so much as a pharmacological tool for FAAH but rather as a template for the development of new therapies simultaneously targeting two proteins, FAAH and TRPV1, the inhibition of which was starting to be seen as potentially very advantageous for the treatment of chronic pain and anxiety. Indeed, AA-5-HT was soon found to behave as a better antihyperalgesic drug against both neuropathic [65,66] and inflammatory [67] pain, and as a stronger anxiolytic compound [68,69], than either selective FAAH or TRPV1 blockers. These promising results prompted the design of other synthetic "dual" FAAH/TRPV1 blockers with a potentially better pharmacokinetic profile (since AA-5-HT can be easily oxidized on both the serotonin and arachidonoyl moieties) [70]. This effort led to the development of OMDM198, which, although less potent than AA-5-HT at two targets, was as efficacious as the serotonin analog at producing antihyperalgesic and anti-inflammatory actions against neuropathic and inflammatory pain [71,72].

Yet another key discovery was the 2012 report that AA-5-HT and other NASs, like other BAFAs, are actually present in mammalian tissues. Verhoeckx et al. [8] discovered that AA-5-HT, *N*-oleoyl-serotonin, *N*-palmitoyl-serotonin, and *N*-stearoyl-serotonin are endogenously present, particularly in the jejunum and ileum of pigs and mice. The authors also observed that NAS formation *in vitro* was stimulated by the addition of serotonin to intestinal tissue incubations and that, in living mice, the pattern of their formation is dependent on the relative amount of fatty acids in the diet. For example, the levels of *N*-docosahexaenoyl-serotonin and *N*-eicosapentaenoyl-serotonin, which, unlike other saturated and monounsaturated NASs, were previously shown to be almost as potent as AA-5-HT at inhibiting FAAH and antagonizing TRPV1 [64], were higher in mice fed with a diet containing fish oil, which is an abundant source of docosahexaenoic and eicosapentaenoic acids. The authors also confirmed FAAH inhibitory activity *in vitro* for most of the NASs and showed that they are able to stimulate GLP-1 secretion from the ileum [8]. Although further experiments aiming at understanding the biosynthetic mechanism of NAS have not yet been performed, a very recent study, apart from demonstrating its presence in bovine and human brain, showed that AA-5-HT can be efficiently oxidized on the 2-position of the indole ring to the corresponding keto-derivative by cytochrome P450 2U1 [73]. This is a very abundant oxygenase in the human thymus, brain, and several other tissues. The ensuing metabolite, 2-oxo-AA-5-HT, was fourfold less active than AA-5-HT at inhibiting FAAH, but was not tested on TRPV1 [73]. By contrast, the fact that AA-5-HT does not increase serotonin levels in the rat brain [53] suggests that it is not easily hydrolyzed by FAAH or other amidases.

5.3.2 Molecular Targets Other than FAAH and TRPV1

Apart from the aforementioned inhibition of FAAH and competitive antagonism of TRPV1, NASs, like NADs, have been also suggested to act as: (1) agonists for GPR119 (in the case of *N*-oleoyl-serotonin [74]), which agrees with the preliminary finding of their stimulatory action on GLP-1 secretion [8]; and (2) inhibitors of T-type Ca^{2+} channels [75], which may underlie some of their analgesic action. Furthermore, AA-5-HT inhibited the release of β-hexosaminidase (IC_{50} = 13.58 μM), a marker of degranulation, and TNF-α (IC_{50} = 12.52 μM), a pro-inflammatory cytokine, in IgE-activated RBL-2H3 cells [76]. Additionally, it suppressed the formation of PGD_2 (IC_{50} = 1.27 μM) and leukotriene B_4 (IC_{50} = 1.2 μM), and the phosphorylation/activation of phospholipase Cγ1/2 and protein kinase Cδ, two enzymes involved in the degranulation process [76]. Since serotonin instead stimulates degranulation, these data indirectly confirm the fact that AA-5-HT is not easily hydrolyzed and that it does not activate 5-HT receptors in mast cells.

5.3.3 *N*-Acyl-Serotonins: Conclusions

Being the latest newcomers to the BAFA family of lipid mediators, NASs are still relatively poorly investigated. In particular, it will be important to understand whether these compounds, like their cognate mediators, NADs and lipoaminoacids, are biosynthesized from the direct condensation of the corresponding fatty acids and amine or via different pathways, and metabolized uniquely via CP450-mediated oxidation. Indeed, a recent study carried out in *Drosophila melanogaster* showed that NASs can be synthesized through the action of an arylalkylamine *N*-acyltransferase [77]. Furthermore, NAS activity as T-type Ca^{2+} channel inhibitors and GPR119 agonists needs to be confirmed and demonstrated to occur also *in vivo*. However, NASs, like NADs, also appear to be present in tissues in low-nanomolar concentrations, thus casting some doubts about the possibility that they interact with targets other than TRPV1 under physiological conditions.

5.4 CONCLUSIONS

We have reviewed here the state of the art of our knowledge on two families of BAFAs, the NADs and NASs, which have the peculiarity of having been first investigated as synthetic compounds and pharmacological tools, and then as putative endogenous mediators. However, still several gaps exist in our understanding of the mechanisms regulating NAD and NAS levels, and this contrasts somehow with the relative abundance of information on their pharmacological actions *in vitro* and *in vivo*, and the promiscuity of their interactions with several potential molecular targets (Figure 5.1). Study of NAD and NAS tissue-level regulation under physiological and pathological conditions will certainly clarify whether these compounds are to be regarded as "trace mediators" of dubious biological importance or as key modulators, like the endocannabinoids and other endocannabinoid-like mediators (see other chapters of this book). This, in turn, will require the development of ever more sensitive analytical techniques for NAD and NAS measurement in tissues and biological fluids (see Chapter 9). It will also require assessment of the levels of these compounds in diets rich or poor in particular polyunsaturated fatty acids, such as linoleic acid or omega-3 fatty acids, which deeply affect other endocannabinoid-like

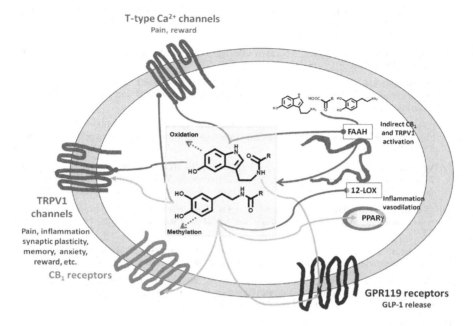

Fig. 5.1. NADs and NASs: possible biosynthetic and degradation pathways and molecular targets with associated physiopathological sequelae. Pointed arrows denote anabolic (full line) and catabolic (broken line) processes or activation of targets. Blunted arrows denote inhibition of targets. Abbreviations: CB1, cannabinoid receptor type-1; FAAH, fatty acid amide hydrolase; GPR119, orphan G-protein-coupled receptor 119; 12-LOX, 12-lipoxygenase; PPARγ, nuclear peroxisome proliferator activated receptor-γ; TRPV1, transient receptor potential vanilloid type-1 (TRPV1) channel.

mediators (see Chapter 2 and Ref. [78] for review and Ref. [79] for an example). These future tasks will increasingly involve and challenge biochemists, pharmacologists, and, in order for this knowledge to be eventually translated to the development of new therapies, nutritionists, and medical doctors.

REFERENCES

[1] Huang SM, Bisogno T, Petros TJ, Chang SY, Zavitsanos PA, Zipkin RE, et al. Identification of a new class of molecules, the arachidonyl amino acids, and characterization of one member that inhibits pain. J Biol Chem 2001;276:42639–44.

[2] Bisogno T, Melck D, Bobrov MYu, Gretskaya NM, Bezuglov VV, De Petrocellis L, et al. *N*-Acyl-dopamines: novel synthetic CB(1) cannabinoid-receptor ligands and inhibitors of anandamide inactivation with cannabimimetic activity *in vitro* and *in vivo*. Biochem J 2000;351(Pt 3):817–24.

[3] Bezuglov V, Bobrov M, Gretskaya N, Gonchar A, Zinchenko G, Melck D, et al. Synthesis and biological evaluation of novel amides of polyunsaturated fatty acids with dopamine. Bioorg Med Chem Lett 2001;11:447–9.

[4] Bisogno T, Melck D, De Petrocellis L, Bobrov MYu, Gretskaya NM, Bezuglov VV, et al. Arachidonoylserotonin and other novel inhibitors of fatty acid amide hydrolase. Biochem Biophys Res Commun 1998;248:515–22.

[5] Bezuglov VV, Bobrov MYu, Archakov AV. Bioactive amides of fatty acids. Biochemistry (Mosc) 1998;63:22–30.

[6] Huang SM, Bisogno T, Trevisani M, Al-Hayani A, De Petrocellis L, Fezza F, et al. An endogenous capsaicin-like substance with high potency at recombinant and native vanilloid VR1 receptors. Proc Natl Acad Sci USA 2002;99:8400–5.

[7] Chu CJ, Huang SM, De Petrocellis L, Bisogno T, Ewing SA, Miller JD, et al. N-Oleoyldopamine, a novel endogenous capsaicin-like lipid that produces hyperalgesia. J Biol Chem 2003;278:13633–9.

[8] Verhoeckx KCM, Voortman T, Balvers MGJ, Hendriks HFJ, Wortelboer HM, Witkamp RF. Presence, formation and putative biological activities of N-acyl serotonins, a novel class of fatty-acid derived mediators, in the intestinal tract. Biochim Biophys Acta 2011;1811:578–86.

[9] Cravatt BF, Giang DK, Mayfield SP, Boger DL, Lerner RA, Gilula NB. Molecular characterization of an enzyme that degrades neuromodulatory fatty-acid amides. Nature 1996;384: 83–7.

[10] Cravatt BF, Prospero-Garcia O, Siuzdak G, Gilula NB, Henriksen SJ, Boger DL, et al. Chemical characterization of a family of brain lipids that induce sleep. Science 1995;268:1506–9.

[11] De Petrocellis L, Chu CJ, Moriello AS, Kellner JC, Walker JM, Di Marzo V. Actions of two naturally occurring saturated N-acyldopamines on transient receptor potential vanilloid 1 (TRPV1) channels. Br J Pharmacol 2004;143:251–6.

[12] De Petrocellis L, Davis JB, Di Marzo V. Palmitoylethanolamide enhances anandamide stimulation of human vanilloid VR1 receptors. FEBS Lett 2001;506:253–6.

[13] Di Marzo V, Melck D, Orlando P, Bisogno T, Zagoory O, Bifulco M, et al. Palmitoylethanolamide inhibits the expression of fatty acid amide hydrolase and enhances the anti-proliferative effect of anandamide in human breast cancer cells. Biochem J 2001;358:249–55.

[14] Ben-Shabat S, Fride E, Sheskin T, Tamiri T, Rhee MH, Vogel Z, et al. An entourage effect: inactive endogenous fatty acid glycerol esters enhance 2-arachidonoyl-glycerol cannabinoid activity. Eur J Pharmacol 1998;353:23–31.

[15] Hu SS, Bradshaw HB, Benton VM, Chen JS, Huang SM, Minassi A, et al. The biosynthesis of N-arachidonoyl dopamine (NADA), a putative endocannabinoid and endovanilloid, via conjugation of arachidonic acid with dopamine. Prostaglandins Leukot Essent Fatty Acids 2009;81:291–301.

[16] Zajac D, Spolnik G, Roszkowski P, Danikiewicz W, Czarnocki Z, Pokorski M. Metabolism of N-acylated-dopamine. PLoS One 2014;9:e85259.

[17] Almási R, Szoke E, Bölcskei K, Varga A, Riedl Z, Sándor Z, et al. Actions of 3-methyl-N-oleoyldopamine, 4-methyl-N-oleoyldopamine and N-oleoylethanolamide on the rat TRPV1 receptor in vitro and in vivo. Life Sci 2008;82:644–51.

[18] Cristino L, Starowicz K, De Petrocellis L, Morishita J, Ueda N, Guglielmotti V, et al. Immunohistochemical localization of anabolic and catabolic enzymes for anandamide and other putative endovanilloids in the hippocampus and cerebellar cortex of the mouse brain. Neuroscience 2008;151:955–68.

[19] Akimov MG, Nazimov IV, Gretskaya NM, Zinchenko GN, Bezuglov VV. Sulfation of N-acyl dopamines in rat tissues. Biochemistry (Mosc) 2009;74:681–5.

[20] Rimmerman N, Bradshaw HB, Basnet A, Tan B, Widlanski TS, Walker JM. Microsomal omega-hydroxylated metabolites of N-arachidonoyl dopamine are active at recombinant human TRPV1 receptors. Prostaglandins Other Lipid Mediat 2009;88:10–7.

[21] Prusakiewicz JJ, Turman MV, Vila A, Ball HL, Al-Mestarihi AH, Di Marzo V, et al. Oxidative metabolism of lipoamino acids and vanilloids by lipoxygenases and cyclooxygenases. Arch Biochem Biophys 2007;464:260–8.

[22] Harrison S, De Petrocellis L, Trevisani M, Benvenuti F, Bifulco M, Geppetti P, et al. Capsaicin-like effects of N-arachidonoyl-dopamine in the isolated guinea pig bronchi and urinary bladder. Eur J Pharmacol 2003;475:107–14.

[23] O'Sullivan SE, Kendall DA, Randall MD. Characterisation of the vasorelaxant properties of the novel endocannabinoid N-arachidonoyl-dopamine (NADA). Br J Pharmacol 2004;141:803–12.

[24] Sagar DR, Smith PA, Millns PJ, Smart D, Kendall DA, Chapman V. TRPV1 and CB(1) receptor-mediated effects of the endovanilloid/endocannabinoid N-arachidonoyl-dopamine on primary afferent fibre and spinal cord neuronal responses in the rat. Eur J Neurosci 2004;20: 175–84.

[25] Marinelli S, Di Marzo V, Florenzano F, Fezza F, Viscomi MT, van der Stelt M, et al. N-Arachidonoyl-dopamine tunes synaptic transmission onto dopaminergic neurons by activating both cannabinoid and vanilloid receptors. Neuropsychopharmacology 2007;32:298–308.

[26] Freestone PS, Guatteo E, Piscitelli F, di Marzo V, Lipski J, Mercuri NB. Glutamate spillover drives endocannabinoid production and inhibits GABAergic transmission in the substantia nigra pars compacta. Neuropharmacology 2014;79:467–75.

[27] Goswami C, Schmidt H, Hucho F. TRPV1 at nerve endings regulates growth cone morphology and movement through cytoskeleton reorganization. FEBS J 2007;274:760–72.

[28] Sharkey KA, Cristino L, Oland LD, Van Sickle MD, Starowicz K, Pittman QJ, et al. Arvanil, anandamide and N-arachidonoyl-dopamine (NADA) inhibit emesis through cannabinoid CB1 and vanilloid TRPV1 receptors in the ferret. Eur J Neurosci 2007;25:2773–82.

[29] Wang Y, Wang DH. Increased depressor response to N-arachidonoyl-dopamine during high salt intake: role of the TRPV1 receptor. J Hypertens 2007;25:2426–33.

[30] Saunders C, Fassett RG, Geraghty DP. Up-regulation of TRPV1 in mononuclear cells of end-stage kidney disease patients increases susceptibility to N-arachidonoyl-dopamine (NADA)-induced cell death. Biochim Biophys Acta 2009;1792:1019–26.

[31] Davies JW, Hainsworth AH, Guerin CJ, Lambert DG. Pharmacology of capsaicin-, anandamide-, and N-arachidonoyl-dopamine-evoked cell death in a homogeneous transient receptor potential vanilloid subtype 1 receptor population. Br J Anaesth 2010;104:596–602.

[32] Farkas I, Tuboly G, Benedek G, Horvath G. The antinociceptive potency of N-arachidonoyl-dopamine (NADA) and its interaction with endomorphin-1 at the spinal level. Pharmacol Biochem Behav 2011;99:731–7.

[33] Szolcsányi J, Sándor Z, Petho G, Varga A, Bölcskei K, Almási R, et al. Direct evidence for activation and desensitization of the capsaicin receptor by N-oleoyldopamine on TRPV1-transfected cell, line in gene deleted mice and in the rat. Neurosci Lett 2004;361:155–8.

[34] Kulisch C, Albrecht D. Effects of single swim stress on changes in TRPV1-mediated plasticity in the amygdala. Behav Brain Res 2013;236:344–9.

[35] Hauer D, Schelling G, Gola H, Campolongo P, Morath J, Roozendaal B, et al. Plasma concentrations of endocannabinoids and related primary fatty acid amides in patients with post-traumatic stress disorder. PLoS One 2013;8:e62741.

[36] Shirazi M, Izadi M, Amin M, Rezvani ME, Roohbakhsh A, Shamsizadeh A. Involvement of central TRPV1 receptors in pentylenetetrazole and amygdala-induced kindling in male rats. Neurol Sci 2014.

[37] Ross HR, Gilmore AJ, Connor M. Inhibition of human recombinant T-type calcium channels by the endocannabinoid N-arachidonoyl dopamine. Br J Pharmacol 2009;156:740–50.

[38] Cazade M, Nuss CE, Bidaud I, Renger JJ, Uebele VN, Lory P, et al. Cross-modulation and molecular interaction at the Cav3.3 protein between the endogenous lipids and the T-type calcium channel antagonist TTA-A2. Mol Pharmacol 2014;85:218–25.

[39] Köfalvi A, Pereira MF, Rebola N, Rodrigues RJ, Oliveira CR, Cunha RA. Anandamide and NADA bi-directionally modulate presynaptic Ca^{2+} levels and transmitter release in the hippocampus. Br J Pharmacol 2007;151:551–63.

[40] De Petrocellis L, Starowicz K, Moriello AS, Vivese M, Orlando P, Di Marzo V. Regulation of transient receptor potential channels of melastatin type 8 (TRPM8): effect of cAMP, cannabinoid CB(1) receptors and endovanilloids. Exp Cell Res 2007;313:1911–20.

[41] Yoo JM, Park ES, Kim MR, Sok DE. Inhibitory effect of N-acyl dopamines on IgE-mediated allergic response in RBL-2H3 cells. Lipids 2013;48:383–93.

[42] Navarrete CM, Pérez M, de Vinuesa AG, Collado JA, Fiebich BL, Calzado MA, et al. Endogenous N-acyl-dopamines induce COX-2 expression in brain endothelial cells by stabilizing mRNA through a p38 dependent pathway. Biochem Pharmacol 2010;79:1805–14.

[43] Navarrete CM, Fiebich BL, de Vinuesa AG, Hess S, de Oliveira AC, Candelario-Jalil E, et al. Opposite effects of anandamide and N-arachidonoyl dopamine in the regulation of prostaglandin E and 8-iso-PGF formation in primary glial cells. J Neurochem 2009;109:452–64.

[44] Sancho R, Macho A, de La Vega L, Calzado MA, Fiebich BL, Appendino G, et al. Immunosuppressive activity of endovanilloids: N-arachidonoyl-dopamine inhibits activation of the NF-kappa B, NFAT, and activator protein 1 signaling pathways. J Immunol 2004;172:2341–51.

[45] McHugh D, Tanner C, Mechoulam R, Pertwee RG, Ross RA. Inhibition of human neutrophil chemotaxis by endogenous cannabinoids and phytocannabinoids: evidence for a site distinct from CB1 and CB2. Mol Pharmacol 2008;73:441–50.

[46] Dang HT, Kang GJ, Yoo ES, Hong J, Choi JS, Kim HS, et al. Evaluation of endogenous fatty acid amides and their synthetic analogues as potential anti-inflammatory leads. Bioorg Med Chem 2011;19:1520–7.

[47] O'Sullivan SE, Kendall DA, Randall MD. Time-dependent vascular effects of endocannabinoids mediated by peroxisome proliferator-activated receptor gamma (PPARγ). PPAR Res 2009;2009:425289.

[48] Wojtalla A, Herweck F, Granzow M, Klein S, Trebicka J, Huss S, et al. The endocannabinoid N-arachidonoyl dopamine (NADA) selectively induces oxidative stress-mediated cell death in hepatic stellate cells but not in hepatocytes. Am J Physiol Gastrointest Liver Physiol 2012;302:G873–87.

[49] Björklund E, Norén E, Nilsson J, Fowler CJ. Inhibition of monoacylglycerol lipase by troglitazone, N-arachidonoyl dopamine and the irreversible inhibitor JZL184: comparison of two different assays. Br J Pharmacol 2010;161:1512–26.

[50] Mahmmoud YA, Gaster M. Uncoupling of sarcoplasmic reticulum Ca^{2+}-ATPase by N-arachidonoyl dopamine. Members of the endocannabinoid family as thermogenic drugs. Br J Pharmacol 2012;166:2060–9.

[51] Chu ZL, Carroll C, Chen R, Alfonso J, Gutierrez V, He H, et al. N-Oleoyldopamine enhances glucose homeostasis through the activation of GPR119. Mol Endocrinol 2010;24:161–70.

[52] Fowler CJ, Tiger G, López-Rodríguez ML, Viso A, Ortega-Gutiérrez S, Ramos JA. Inhibition of fatty acid amidohydrolase, the enzyme responsible for the metabolism of the endocannabinoid anandamide, by analogues of arachidonoyl-serotonin. J Enzyme Inhib Med Chem 2003;18:225–31.

[53] de Lago E, Petrosino S, Valenti M, Morera E, Ortega-Gutierrez S, Fernandez-Ruiz J, et al. Effect of repeated systemic administration of selective inhibitors of endocannabinoid inactivation on rat brain endocannabinoid levels. Biochem Pharmacol 2005;70:446–52.

[54] Suplita, 2nd RL, Farthing JN, Gutierrez T, Hohmann AG. Inhibition of fatty-acid amide hydrolase enhances cannabinoid stress-induced analgesia: sites of action in the dorsolateral periaqueductal gray and rostral ventromedial medulla. Neuropharmacology 2005;49:1201–9.

[55] Soria-Gómez E, Matias I, Rueda-Orozco PE, Cisneros M, Petrosino S, Navarro L, et al. Pharmacological enhancement of the endocannabinoid system in the nucleus accumbens shell stimulates food intake and increases c-Fos expression in the hypothalamus. Br J Pharmacol 2007;151:1109–16.

[56] Darmani NA, McClanahan BA, Trinh C, Petrosino S, Valenti M, Di Marzo V. Cisplatin increases brain 2-arachidonoylglycerol (2-AG) and concomitantly reduces intestinal 2-AG and anandamide levels in the least shrew. Neuropharmacology 2005;49:502–13.

[57] Llorente R, Llorente-Berzal A, Petrosino S, Marco EM, Guaza C, Prada C, et al. Gender-dependent cellular and biochemical effects of maternal deprivation on the hippocampus of neonatal rats: a possible role for the endocannabinoid system. Dev Neurobiol 2008;68: 1334–47.

[58] López-Gallardo M, Llorente R, Llorente-Berzal A, Marco EM, Prada C, Di Marzo V, et al. Neuronal and glial alterations in the cerebellar cortex of maternally deprived rats: gender differences and modulatory effects of two inhibitors of endocannabinoid inactivation. Dev Neurobiol 2008;68:1429–40.

[59] Di Marzo V, Capasso R, Matias I, Aviello G, Petrosino S, Borrelli F, et al. The role of endocannabinoids in the regulation of gastric emptying: alterations in mice fed a high-fat diet. Br J Pharmacol 2008;153:1272–80.

[60] De Filippis D, D'Amico A, Cipriano M, Petrosino S, Orlando P, Di Marzo V, et al. Levels of endocannabinoids and palmitoylethanolamide and their pharmacological manipulation in chronic granulomatous inflammation in rats. Pharmacol Res 2010;61:321–8.

[61] Bifulco M, Laezza C, Valenti M, Ligresti A, Portella G, Di Marzo V. A new strategy to block tumor growth by inhibiting endocannabinoid inactivation. FASEB J 2004;18:1606–8.

[62] Ligresti A, Bisogno T, Matias I, De Petrocellis L, Cascio MG, Cosenza V, et al. Possible endocannabinoid control of colorectal cancer growth. Gastroenterology 2003;125:677–8.

[63] D'Argenio G, Valenti M, Scaglione G, Cosenza V, Sorrentini I, Di Marzo V. Up-regulation of anandamide levels as an endogenous mechanism and a pharmacological strategy to limit colon inflammation. FASEB J 2006;20:568–70.

[64] Ortar G, Cascio MG, De Petrocellis L, Morera E, Rossi F, Schiano-Moriello A, et al. New *N*-arachidonoylserotonin analogues with potential "dual" mechanism of action against pain. J Med Chem 2007;50:6554–69.

[65] Maione S, De Petrocellis L, de Novellis V, Moriello AS, Petrosino S, Palazzo E, et al. Analgesic actions of *N*-arachidonoyl-serotonin, a fatty acid amide hydrolase inhibitor with antagonistic activity at vanilloid TRPV1 receptors. Br J Pharmacol 2007;150:766–81.

[66] de Novellis V, Vita D, Gatta L, Luongo L, Bellini G, De Chiaro M, et al. The blockade of the transient receptor potential vanilloid type 1 and fatty acid amide hydrolase decreases symptoms and central sequelae in the medial prefrontal cortex of neuropathic rats. Mol Pain 2011;7:7.

[67] Costa B, Bettoni I, Petrosino S, Comelli F, Giagnoni G, Di Marzo V. The dual fatty acid amide hydrolase/TRPV1 blocker, *N*-arachidonoyl-serotonin, relieves carrageenan-induced inflammation and hyperalgesia in mice. Pharmacol Res 2010;61:537–46.

[68] Micale V, Cristino L, Tamburella A, Petrosino S, Leggio GM, Drago F, et al. Anxiolytic effects in mice of a dual blocker of fatty acid amide hydrolase and transient receptor potential vanilloid type-1 channels. Neuropsychopharmacology 2009;34:593–606.

[69] John CS, Currie PJ. *N*-Arachidonoyl-serotonin in the basolateral amygdala increases anxiolytic behavior in the elevated plus maze. Behav Brain Res 2012;233:382–8.

[70] Morera E, De Petrocellis L, Morera L, Moriello AS, Ligresti A, Nalli M, et al. Synthesis and biological evaluation of piperazinyl carbamates and ureas as fatty acid amide hydrolase (faah) and transient receptor potential (trp) channel dual ligands. Bioorg Med Chem Lett 2009;19:6806–9.

[71] Maione S, Costa B, Piscitelli F, Morera E, De Chiaro M, Comelli F, et al. Piperazinyl carbamate fatty acid amide hydrolase inhibitors and transient receptor potential channel modulators as "dual-target" analgesics. Pharmacol Res 2013;76:98–105.

[72] Starowicz K, Malek N, Mrugala M, Makuch W, Comelli F, Ortar G, et al. OMDM198, a compound targeting both TRPV1 and fatty acid amide hydrolase: a new pain management strategy in osteoarthritis? In: 22nd Annual Symposium of the International Cannabinoid Research Society, Freiburg, Germany July 22–27, 2012, p. 34.

[73] Siller M, Goyal S, Yoshimoto FK, Xiao Y, Wei S, Guengerich FP. Oxidation of endogenous N-arachidonoylserotonin by human cytochrome P450 2U1. J Biol Chem 2014;289:10476–87.

[74] Syed SK, Bui HH, Beavers LS, Farb TB, Ficorilli J, Chesterfield AK, et al. Regulation of GPR119 receptor activity with endocannabinoid-like lipids. Am J Physiol Endocrinol Metab 2012;303:E1469–78.

[75] Cazade M, Nuss CE, Bidaud I, Renger JJ, Uebele VN, Lory P, et al. Cross-modulation and molecular interaction at the Cav3.3 protein between the endogenous lipids and the T-type calcium channel antagonist TTA-A2. Mol Pharmacol 2014;85:218–25.

[76] Yoo JM, Sok DE, Kim MR. Effect of endocannabinoids on IgE-mediated allergic response in RBL-2H3 cells. Int Immunopharmacol 2013;17:123–31.

[77] Dempsey DR, Jeffries KA, Anderson RL, Carpenter AM, Rodriquez Opsina S, Merkler DJ Identification of an arylalkylamine N-acyltransferase from *Drosophila melanogaster* that catalyzes the formation of long-chain N-acylserotonins. FEBS Lett. 2014;588:594–9.

[78] Banni S, Di Marzo V. Effect of dietary fat on endocannabinoids and related mediators: consequences on energy homeostasis, inflammation and mood. Mol Nutr Food Res 2010;54:82–92.

[79] Alvheim AR, Malde MK, Osei-Hyiaman D, Lin YH, Pawlosky RJ, Madsen L, et al. Dietary linoleic acid elevates endogenous 2-AG and anandamide and induces obesity. Obesity (Silver Spring) 2012;20:1984–94.

CHAPTER 6

The Pharmacology of Prostaglandin $F_{2\alpha}$ Ethanolamide and Bimatoprost Reveals a Unique Feedback Mechanism on Endocannabinoid Actions

David F. Woodward, Jenny W. Wang

6.1 INTRODUCTION

Since the 1997 discovery that the mammalian endocannabinoid anandamide could be converted to prostaglandin E_2-ethanolamide (prostamides E_2) by cyclooxygenase-2 (COX-2) [1], the biological and therapeutic significance of this has been gradually appreciated. Bimatoprost, being a commercially available drug, has provided much of the impetus for the early research advances. Bimatoprost was found to be functionally distinct from its free acid congeners prostaglandin $F_{2\alpha}$ ($PGF_{2\alpha}$) and 17-phenyl $PGF_{2\alpha}$ [2,3]. The development of prostamide antagonists [4] and the molecular characterization of bimatoprost-sensitive receptors [5] allowed bimatoprost to be described as pharmacologically distinct from prostanoids FP receptor agonists. Prostamide $F_{2\alpha}$ was found as the naturally occurring equivalent of bimatoprost [6,7]. Since prostamides E_2 may be biosynthesized from anandamide by COX-2 [1], it was not surprising that prostamides $F_{2\alpha}$ could be similarly biosynthesized [8].

The Endocannabinoidome: The World of Endocannabinoids and Related Mediators. DOI: 10.1016/B978-0-12-420126-2.00006-7
Copyright © 2015 Elsevier Inc. All rights reserved

The prostamide endoperoxides formed by COX-2 mediated oxygenation of anandamide may be subsequently reduced to prostamides $F_{2\alpha}$ by two quite different PGF synthases. These reductases are AKR1C3/prostaglandin F synthase [9] and aldo-keto reductase (AKR) and prostamide/prostaglandin F synthase, which belongs to the thioredoxin-like superfamily [10].

Prostamides D_2, E_2, and I_2 are also formed from anandamide [8], but detailed analysis of these biosynthetic routes remains to be undertaken. The functional significance of endogenous prostamides has only recently been studied and much more is required. This will be discussed, in addition to reviewing known activities of bimatoprost and exogenously administered prostamides. Expanding knowledge with respect to the therapeutic utility of the prostamides forms the final theme of this review.

6.2 PHARMACOLOGY

The pharmacology of prostamides $F_{2\alpha}$ and its analog bimatoprost have been extensively reviewed on many occasions and, therefore, this topic will be only briefly summarized. The central issue was the pharmacological relationship of prostamides $F_{2\alpha}$ and bimatoprost to $PGF_{2\alpha}$, 17-phenyl $PGF_{2\alpha}$, and other FP receptor agonists. Comparative agonist studies revealed that neutral prostamides $F_{2\alpha}$ and its synthetic analog could be essentially equipotent to $PGF_{2\alpha}$ and FP receptor agonists in some pharmacological preparations, whereas in many FP receptor preparations the prostamides exhibited little activity [11,12]. In dispense feline iris smooth muscle cells, bimatoprost stimulated a different cell population that is responsive to $PGF_{2\alpha}$ and 17-phenyl $PGF_{2\alpha}$ [3]. This was considered to provide compelling evidence for pharmacological differentiation of prostamides from prostanoid FP receptor agonists. The next research phase was the attempted discovery of prostamide antagonists, which proved a successful endeavor. The prototype antagonists AGN-204396 and -204397 blocked the effects of bimatoprost and prostamides $F_{2\alpha}$ in the feline iris preparations but not those of $PGF_{2\alpha}$ and FP receptor agonists [4,11]. Prostamide antagonists 100 times more potent were subsequently developed [5,11,13]. These antagonists, AGN-211334, 5, and 6, were sufficiently potent to be used in living animal studies. Thus, AGN-211336 was shown to attenuate bimatoprost-induced ocular hypotension in dogs but not the ocular hypotensive effects of the FP agonist prodrug latanoprost [14]. This experiment provided definitive evidence that the ocular hypotensive effects of bimatoprost were not prostanoid FP receptor mediated. These findings are supported by clinical studies of an alternative design. In glaucoma patients refractory to latanoprost therapy, bimatoprost was fully efficacious [15–17]. This experimental outcome is not consistent with both drugs interacting with the same target receptor.

The discovery of a number of FP receptor mRNA splicing variants with desensitization properties different from those displayed by wild-type FP receptors [5,18–20]. These FP receptor variants were not, however, bimatoprost sensitive. In an attempt to model the prostamides receptor, certain splicing variants and the wild-type FP receptor were cotransfected and responses to bimatoprost and $PGF_{2\alpha}$ were evaluated [5]. This experimental approach provided a bimatoprost-sensitive pharmacological replicate when the splicing variant designated altFP4 was cotransfected with the wild-type FP receptor. Bimatoprost, but not $PGF_{2\alpha}$, induced Cyr 61 upregulation was blocked by the prostamides antagonist AGN-211335 [5]. In the altFP4/FP cotransfectants, bimatoprost evoked a transient Ca^{2+} signal followed by an oscillating Ca^{2+} signal of low frequency. The secondary oscillating signal was susceptible to antagonism by AGN-211335 [5]. The molecular characterization of prostamide receptors has not yet advanced from this point. Various permutations of FP variant and wild-type receptors should be undertaken and pharmacologically characterized.

6.3 ENDOGENOUS PROSTAMIDE $PGF_{2\alpha}$

Anandamide oxygenation by COX-2 was originally demonstrated using purified human enzyme and HFF cells [1]. This finding was confirmed and greatly expanded by detailed studies on the fate of exogenously administered endocannabinoids [8]. Studies on the human colon adenocarcinoma cells line HCA-7, which constitutively express COX-2, demonstrated in the same way that PGE_2 and $PGF_{2\alpha}$ are formed following arachidonic acid treatment, prostamides E_2 and $F_{2\alpha}$ are biosynthesized after anandamide addition to the cells [8]. The downstream formation of prostamide $F_{2\alpha}$ was also studied and found to potentially involve two distinct reductase enzymes [21,22]. Prostaglandin F synthase (PGF synthase), a member of the AKR superfamily and designated AKR1C3 for the human isoform [23], was found to produce prostamide $F_{2\alpha}$ from the endoperoxide intermediate and 11β-prostamide $F_{2\alpha}$ from PGD_2 [21]. The subsequent discovery of a new enzyme prostamide/prostaglandin F synthase [22], a member of the thioredoxin-like superfamily, was invaluable and instructional as an enzyme that specifically converted the acidic and amido H_2 endoperoxides to $PGF_{2\alpha}$ and prostamide $F_{2\alpha}$, respectively. The distribution of these enzymes was also quite different: PGF synthase expression was prominent in the lung and liver, whereas prostamide/PGF synthase was highly expressed in neuronal tissues [21,22]. It can be concluded from the sum total of these studies that, given an adequate supply of anandamide substrate, the biosynthetic machinery to biosynthesize prostamide $F_{2\alpha}$ exists. The next self-interrogative issue raised was the physiological significance of this pathway, which would preferentially require demonstration of prostamide $F_{2\alpha}$ biosynthesis from endogenously present anandamide. An *in vitro* study on LPS-treated human macrophages identified low levels of a molecule characteristic of prostamide $F_{2\alpha}$ (Wang et al., unpublished data), but this finding was inconsistent and close to the detection limitations.

Other studies to detect endogenous prostamide $F_{2\alpha}$ were conducted on tissues. Studies on tissues obtained from in-life sources present greater challenges. First, tissue extraction is an integral part of the process but may be performed at greatly reduced efficiency compared to analyzing cell medium. Second, and more importantly, is the potential for a confounding dilution factor. For example, in tissue extracts detection of an important local neurotransmitter, located in a small but important neuronal plexus in a large tissue, may be compromised or even remain undetected. Time from tissue excision to extraction could also introduce unreliability. Endogenous prostamide $F_{2\alpha}$ detection was attempted in ocular tissue obtained from human donors but results were sporadic and convincing in only a few instances (Wang et al., unpublished results). An *in vivo* study was performed in mice comparing fatty acid amide hydrolase (FAAH) enzyme gene deleted mice, to attenuate anandamide hydrolysis with wild-type animals [7]. These studies were performed with or without exogenous anandamide administration [7]. Prostamide $F_{2\alpha}$ was detected only in FAAH gene deleted mice that received exogenous anandamide [7]. These studies predated the discovery of prostamide/

PGF synthase [22] and such mouse studies were designed according to the contemporaneous premise that PGF synthase [2,24] would be the enzymatic source of prostamide F$_{2\alpha}$. Thus, the FAAH knock-out mice study [7] was designed to investigate prostamide F$_{2\alpha}$ levels in tissues where they were thought most likely to be present; at that time where PGF synthase was most abundantly expressed. Nonetheless, the FAAH knock-out mice study did achieve one useful outcome. Living animals could biosynthesize prostamide F$_{2\alpha}$, albeit under somewhat contrived circumstances. The discovery of an enzyme, essentially dedicated to the biosynthesis of PGF$_{2\alpha}$ and prostamide F$_{2\alpha}$ [22], changed the landscape of such studies by signposting appropriate tissues for study. Prostamide/PGF synthase was found abundantly expressed in spinal cord and brain [22,25], which suggested central nervous system (CNS) tissue as a likely site for the detection of endogenously synthesized prostamide F$_{2\alpha}$. This proved to be the case. Studies on the rat spinal cord revealed levels of endogenous prostamide F$_{2\alpha}$ sufficient to be measured as changed in response to inflammatory stimuli [26]. This represents the first evidence that prostamide F$_{2\alpha}$ exists endogenously and exerts biological function.

6.4 PROSTAMIDE RECEPTOR DETECTION

The bimatoprost-sensitive prostamide receptor has been modeled by co-transfecting the wild-type FP receptor with an alternative mRNA splicing variant, specifically altFP4 [5]. Given the technical difficulties associated with receptor protein detection of an FP/altFP4 heterodimeric or other complex, receptor expression has relied on detection of transcripts. Typically, identification has been determined after functional responses to bimatoprost have been established [27,28]. In both cases [27,28] the truncated mRNA splicing variant, altFP$_4$, was detected along with wild-type FP receptor gene expression. In the case of the hair follicle studies [27], the altFP1 variant was also found in the dermal papilla and the surrounding connective tissue sheath. However, its role in bimatoprost-induced hypertrichosis remains undetermined. Primers common for all FP variants have been used in studies on human adipocytes [28] and the human ciliary body [5], which is a major target tissue for bimatoprost in glaucoma therapy [29]. It goes without saying that whenever *PTGFR* (FP) receptor gene expression is detected, there is also a possibility that its various FP splicing variants [18–20,30] will be detected. With the exception of altFP4, the function of the various FP variants, if any, remains to be investigated. It remains unknown whether these other variants will physically interact to form functionally viable receptor complexes that would respond to bimatoprost and other electrochemically neutral $PGF_{2\alpha}$ molecular species.

6.5 BIOLOGICAL FUNCTION AND THERAPEUTICS

One of the most significant recent findings pertaining to the role of electrochemically neutral PGs (PG glyceryl esters and ethanolamides) and endocannabinoids is the discovery of substrate-specific inhibition of endocannabinoid oxygenation by COX-2 [31]. This substrate-specific inhibition is achieved with the (R)-enantiomers of the profens; specifically the (R)-isomers of flurbiprofen, naproxen, and, ibuprofen [31]. These compounds behave as weak, competitive inhibitors of arachidonic acid oxygenation but are potent inhibitors of endocannabinoid oxygenation. The importance of these findings is reflected in a previous living animal study where (R)-flurbiprofen restored endogenous endocannabinoids to reduce neuropathic pain in rodents [32]. In sciatic nerve injury models, (R)-flurbiprofen reduced glutamate release in the dorsal horn and thereby attenuated behavior indicative of neuropathic pain [32]. The antinociceptive response was attributed to restoration of diminished endocannabinoid levels, an imbalance being created by axonal injury. Considering the imbalance in a more detailed manner, and in cells and neuronal tissue, anandamide levels and those of its entourage congeners oleoyl- and palmitoyl-ethanolamide appeared the most affected by (R)-flurbiprofen [32]. Both 2- and 1-arachidonyl glyceryl ester species appeared little affected and were at surprisingly low levels in the dorsal root ganglia, dorsal horn, and the forebrain/cortex. These findings elevate anandamide and its COX-2 oxygenation products to a potentially more prominent position as central nociceptive mediators.

To date prostamides D_2 and E_2 have not been studied as nociceptive stimuli. In marked contrast, prostamide$F_{2\alpha}$ has been the subject of a detailed and extensive study [26]. This was likely prompted by reports on the abundant expression of the key biosynthetic enzyme, prostamide/PGF synthase, in central neuronal tissues [22,25]. A stated objective of this study was to investigate the possible explanations for disappointing clinical results with the FAAH inhibitor PF-04457845 [26,33]. Attempting to experimentally model the clinical trial on PF-04457845 on osteoarthritis, kaolin and carrageenin were sequentially injected into the knee joint of rats [26]. Concentrations of anandamide and 2-AG in the spinal cord were unchanged by modeling knee joint inflammation. Nevertheless, prostamide $F_{2\alpha}$ was endogenously present in the spinal cord and its level was elevated in response to inflammation [26]. Results obtained with prostamide $F_{2\alpha}$ were consistent with a role as a nociceptive mediator, as follows (i) prostamide $F_{2\alpha}$ levels were increased in response to articular inflammation; (ii) increased prostamide $F_{2\alpha}$ levels were normalized by inhibition of COX enzymes; (iii) prostamide $F_{2\alpha}$ increased firing of nociceptive neurons; (iv) prostamide $F_{2\alpha}$-induced nociception was blocked by the prostamide antagonist AGN-211336 but not by the prostanoid FP receptor antagonist AL-8810. Prostamide E_2, PGE_2-glyceryl ester, and $PGF_{2\alpha}$-glyceryl ester, were much less abundant (<1 pmol/g tissue) than prostamide $F_{2\alpha}$ [26]. Taken together, these results clearly indicate prostamide $F_{2\alpha}$ as a nociceptive mediator.

Although prostamide $F_{2\alpha}$ may be a significant participant in pain transmission where a prostamide F synthesizing enzyme is highly expressed, in other CNS and peripheral nervous system (PNS) regions different prostamides may assume greater importance depending on the PG synthases expressed. For the purpose of investigating the potential role of prostamides in other experimental pain models, the effect of a prostamide antagonist on capsaicin-induced ocular surface nociception, a pain model likely to display a prostanoid-mediated component [34,35], was employed. The corneal nociceptive responses to capsaicin in rats were significantly reduced by a prostamide antagonist (Figure 6.1). In this experimental setting, prostanoid $F_{2\alpha}$ is unlikely to be the nociceptive prostamide for two reasons: (i) prostamide E_2 is the only detectable prostamide in the cornea [36]; and (ii) bimatoprost, a prostamide $F_{2\alpha}$ analog, is widely used as a topical medication, but ocular pain or discomfort is rarely reported.

The interrelationship between the endocannabinoids anandamide and 2-AG and their respective COX-2 metabolites, the prostamides and PG-glyceryl esters, in regulating pain neurotransmission is not becoming less complicated. Unifying hypotheses have been proposed but

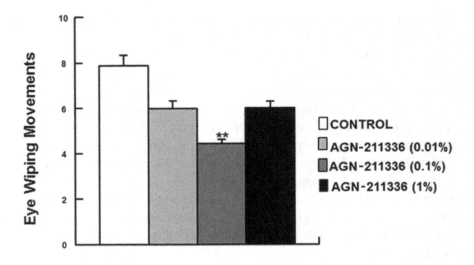

Fig. 6.1. The effect of the prostamide antagonist AGN-211336 on 10 μg/mL capsaicin induced ocular surface pain/discomfort in rats. Thirty minute-pretreatment of AGN-211336 at 0.01% (n = 24), 0.1% (n = 24), and 1% (n = 36) were compared to vehicle control group (n = 35). P < 0.01 is denoted by **, according to ANOVA analysis compared to vehicle control.

even the most recent [12] considers only scenarios where the endocannabinoids and their oxygenation products are in balance. It is becoming clear that basal anandamide and/or 2-AG levels may be essentially unchanged in some pain models [26,32], certainly in models where spinal cord neurotransmission is a subject of investigation. Rather than a singular hypothesis, a more satisfactory explanation would perhaps involve consideration of tissue-dependent situations. Considering COX-2 as a pivotal enzyme, the following scenarios are diagrammatically postulated (Figure 6.2). This theoretical schematic does account for enzymatic hydrolysis of 2-AG by MAG-lipase as a source of PGs in the CNS [37], where a more complex schematic would be necessary.

In the CNS, there is no proof for a straightforward mechanism where anandamide and prostamide $F_{2\alpha}$ exert opposing effects and their relative local concentrations define neurotransmission. This mechanistic relationship between anandamide and prostamide $F_{2\alpha}$ is, however, operative in regulating adipogenesis [28]. Thus, prostamide $F_{2\alpha}$ negatively regulates adipogenesis and anandamide is proadipogenic. Prostamide $F_{2\alpha}$ inhibits the early differentiation of preadipocytes into adipocytes and provides a feed-forward mechanism by upregulation of COX-2 and PTGFR

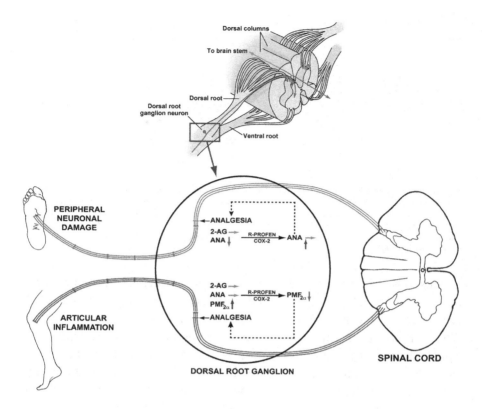

Fig. 6.2. *Peripheral or axonal damage to efferent neurons produces a decrease in anandamide and certain entourage endocannabinoids. The substrate-specific COX-2 inhibitor (R)-flurbiprofen restores anandamide levels, with a concomitant CB receptor–mediated analgesic response: 2-AG levels remain unaltered [31,32]; articular inflammation does not alter endocannabinoid levels but initiates prostamide F$_{2\alpha}$ biosynthesis in the spinal cord, which is involved in neurotransmission of the nociceptive response [26].*

products, which are downregulated during adipogenesis, to levels much higher than those normally observed in preadipocytes. This regulates the amplification provided by increased anandamide biosynthesis, CB$_1$ receptor upregulation, and PPARγ activation [28]. Again, the importance of endocannabinoid oxygenation by COX-2 in preventing preadipocyte differentiation was proven by using the substrate-selective COX-2 inhibitor R-flurbiprofen [28]. Differentiation of preadipocytes into adipocytes was enhanced by R-flurbiprofen treatment. The preadipocyte to adipocyte regulatory mechanisms are depicted diagrammatically in Figure 6.3. It should be noted that a comparable regulatory mechanism for adipogenesis does not exist for PGF$_{2\alpha}$ and arachidonic acid

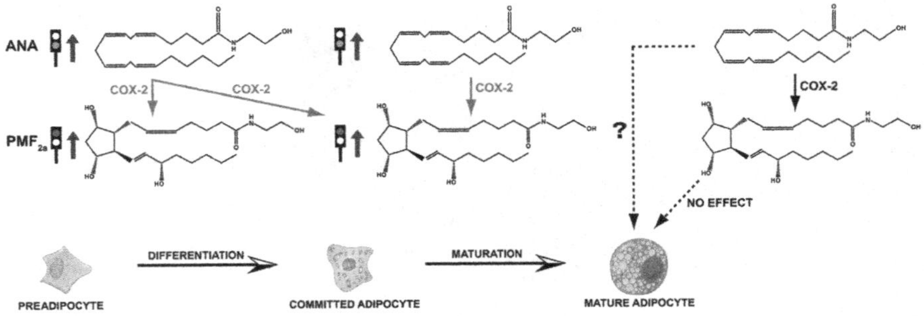

Fig. 6.3. Regulation of adipogenesis involves a feed-forward mechanism whereby prostamide $F_{2\alpha}$, a COX-2 product of anandamide, negatively regulates preadipocyte differentiation. This opposes proadipogenic, CB_1 receptor–mediated signals produced by anandamide biosynthesized by preadipocytes and immature adipocytes [28]. This mechanism may not be operative at the terminally differentiated mature adipocyte level [38]. Negative regulation is depicted as red arrows and traffic signals; positive regulation, provided by increased anandamide levels, is indicated by green traffic signals.

because arachidonic acid does not possess a direct, receptor-mediated activity comparable that exhibited by anadamide at cannabinoid (CB) receptors. A recent but crude study on mature adipocytes derived from 3T3-L1 cells showed no effect for bimatoprost on intracellular lipid volume, although FP-receptor activation clearly showed a reduction [38]. This result would support the view that bimatoprost effects are restricted to the earlier phases of adipogenesis but this interpretation does not take into account the physicochemical properties of bimatoprost versus those of the corresponding prostanoic acids [38]. Detergent molecules are currently being developed for local reduction of adipose tissue and, since the FP agonists used would be weak detergent-like molecules and bimatoprost is electrochemically neutral, the differential results obtained with FP agonists versus bimatoprost may simply reflect detergent-induced lipolysis.

Although the most recent evidence [28] has provided detailed insight into the mechanism of action of bimatoprost as an antiadipogenic agent, the effects of bimatoprost on fat deposition were first observed in periorbital tissue. The periorbital changes were originally described as a general appearance of "sunken eyes," where the size of the fat pads was reduced [39–42]. An *in vitro* study on human orbital preadipocytes found that bimatoprost produced the most significant reduction in adipogenesis compared to $PGF_{2\alpha}$ and the antiglaucoma FP receptor agonists latanoprost, travoprost, and tafluprost [43]. Bimatoprost exhibited consistent superiority on all parameters studied; fat accumulation, lipoprotein lipase (LPL) expression, CCAAT-enhancer binding protein α (C/EBPα) expression, and peroxisome proliferator-activated receptor γ (PPARγ) mRNA expression. The effects of prostamide F$_{2\alpha}$ and its relation to anandamide in periorbital adipogenesis were not studied [43].

The effects of bimatoprost and the prostamides in glaucoma and hair growth have been recently and extensively reviewed [12]. Nevertheless, there is new information on glaucoma, where bimatoprost has been used as first line antiglaucoma therapy for many years. The most recent advancement is the use of a 0.01% dose, which has essentially eliminated the ocular surface hyperemic effect produced in a significant percentage of patients by the original 0.03% once daily dosing regimen [44–47]. Bimatoprost (Lumigan™) is essentially equivalent in lowering intraocular pressure at 0.01 or 0.03% doses, but the incidence and severity of ocular surface hyperemia are greatly reduced [45–48]. In observation studies intended to reflect normal clinical practice, 93.9, 93.2, or 93.5% of patients reported no adverse events including ocular and conjunctival hyperemia [46,47]. The absence of ocular surface redness as a prominent side effect of bimatoprost at a 0.01% dose is consistent with the pharmacological profile of bimatoprost. Thus, although PGF$_{2\alpha}$, latanoprost free acid, and other prostanoid FP receptor agonists potently produce NO-endothelium dependent vasorelaxation, bimatoprost has only a weak effect [48,49]. A combination of bimatoprost and timolol results in increased ocular hypotensive efficacy but the ocular surface redness produced by 0.03% bimatoprost appears particularly susceptible to reduction by timolol addition compared to other prostanoid/timolol combinations [50,51]. Although the precise underlying mechanism is not established, enhanced α_1-adrenergic tone and reduced NO production have been suggested [50].

The neuroprotective effects of bimatoprost are now becoming a topic of interest. Following a study on bimatoprost and several prostanoid FP receptor agonist on rat retinal ganglion cell survival *in vitro* [52], bimatoprost was specifically studied *in vivo* [53]. This study provided the first evidence for a retinal neuroprotective effect for bimatoprost in living animals [53]. Further *in vitro* studies on bimatoprost showed a direct neuroprotective effect against oxidative stress, glutamate, and serum deprivation–induced retinal ganglion cell death, although 17-phenyl

$PGF_{2\alpha}$ was reported as more active [53]. At the signal transduction level, bimatoprost-induced activation of Akt and ERK and a PI3-kinase inhibitor attenuated the neuroprotective effects of bimatoprost [53]. The discovery of prostamide $F_{2\alpha}$ as a spinal neurotransmitter [26] indicates potential medical utility for a prostamide antagonist, for example, AGN-211335 as an analgesic therapeutic.

Although bimatoprost, marketed as Latisse™, is a hair growth therapy, nothing of significance has recently been reported since this subject was last reviewed [12]. Briefly, bimatoprost, as the intact molecule, stimulates murine pelage hair growth [54] and hair growth in human isolated scalp hair follicles [27]. Recently, bimatoprost was found to be osteogeneic in *Danio rerio* [28]. In most cases, therapeutic modalities based on prostamide $F_{2\alpha}$ would comprise an agonist mimetic such as bimatoprost. The biological effects of bimatoprost and prostamide $F_{2\alpha}$ continue to diversify and expand.

ACKNOWLEDGMENT

The authors thank Dr. Clayton Spada for creating the figures.

REFERENCES

[1] Yu M, Ives D, Ramesha CS. Synthesis of prostaglandin E2 ethanolamide from anandamide by cyclooxygenase-2. J Biol Chem 1997;272:21181–6.

[2] Woodward DF, Krauss AH-P, Chen J, Liang Y, Li C, Protzman CE, et al. Pharmacological characterization of a novel antiglaucoma agent, bimatoprost (AGN 192024). J Pharmacol Exp Ther 2003;305:772–85.

[3] Spada CS, Krauss AH-P, Woodward DF, Chen J, Protzman CE, Nieves AL, et al. Bimatoprost and prostaglandin F2α selectively stimulate intracellular calcium signaling in different cat iris sphincter cells. Exp Eye Res 2005;80:135–45.

[4] Woodward DF, Krauss AH-P, Wang JW, Protzman CE, Nieves AL, Liang Y, et al. Identification of an antagonist that selectively blocks the activity of prostamides (prostaglandin-ethanolamides) in the feline iris. Br J Pharmacol 2007;150:342–52.

[5] Liang Y, Woodward DF, Guzman VM, Li C, Scott DF, Wang JW, et al. Identification and pharmacological characterization of the prostaglandin FP receptor and FP receptor variant complexes. Br J Pharmacol 2008;154:1079–93.

[6] Matias I, Chen J, De Petrocellis L, Bisogno T, Ligresti A, Fezza F, et al. Prostaglandin ethanolamides (prostamides): *in vitro* pharmacology and metabolism. J Pharmacol Exp Ther 2004;309:745–57.

[7] Weber A, Ni J, Ling K-HJ, Acheampong A, Tang-Liu DD-S, Cravatt BF, et al. Formation of prostaglandin 1-ethanolamides (prostamides) from anandamide in fatty acid amide hydrolase knockout (FAAH−/−) mice analyzed by high performance liquid chromatography with tandem mass spectrometry. J Lipid Res 2004;45:757–63.

[8] Kozak KR, Crews BC, Morrow JD, Wang LH, Ma YH, Weinander R, et al. Metabolism of the endocannabinoids, 2-arachidonylglycerol and anandamide, into prostaglandin, thromboxane and prostacyclin glycerol esters and ethanolamides. J Biol Chem 2002;277:44877–85.

[9] Koda N, Tsutsui Y, Niwa H, Ito S, Woodward DF, Watanabe K. Synthesis of prostaglandin F ethanolamide by prostaglandin ethanolamide by prostaglandin F synthase and identification of bimatoprost as a potent inhibitor of the enzyme: new enzyme method using LC/ESI/MS. Arch Biochem Biophys 2004;424:128–36.

[10] Moriuchi H, Koda N, Okuda-Ashitaka E, Daiyasu H, Ogasawara K, Toh H, et al. Molecular characterization of a novel type of prostamide/prostaglandin F synthase, belonging to the thioredoxin-like superfamily. J Biol Chem 2008;283:792–801.

[11] Woodward DF, Carling RW, Cornell CL, Fliri HG, Martos JL, Pettit SN, et al. The pharmacology and therapeutic relevance of endocannabinoid derived cyclo-oxygenase (COX)-2 products. Pharmacol Ther 2008;120:71–80.

[12] Woodward DF, Wang JW, Poloso NJ. Recent progress in prostaglandin F2α ethanolamide (prostamide F2α) research and therapeutics. Pharmacol Rev 2013;65:1135–47.

[13] Wan Z, Woodward DF, Cornell CL, Fliri HG, Martos JL, Pettit SN, et al. Bimatoprost, prostamide activity, and conventional drainage. Invest Ophthalmol Vis Sci 2007;4:4107–15.

[14] Woodward DF, Liang Y, Wang JW, Li C, Guzman VM, Kharlamb AB, et al. Pharmacological differentiation of bimatoprost and latanoprost induced ocular hypotension by a second generation prostamide antagonist (AGN 211336). In: Goodwin GM editor. Prostaglandins: biochemistry, functions, types, and roles. Hauppauge, NY: Nova Science Publishers Inc.; 2009. p. 272–9.

[15] Williams RD. Efficacy of bimatoprost in glaucoma and ocular hypertension unresponsive to latanoprost. Adv Ther 2002;19:275–81.

[16] Gandolfi SA, Cimino L. Effect of bimatoprost on patients with primary open-angle glaucoma or ocular hypertension who are nonresponders to latanoprost. Ophthalmology 2003;110: 609–14.

[17] Sonty S, Donthamsetti V, Vangipuram G, Ahmad A. Long-term IOP lowering with bimatoprost in open-angle glaucoma patients poorly responsive to latanoprost. J Ocular Pharmacol Ther 2008;24:517–20.

[18] Pierce KL, Bailey TJ, Hoyer P, Gil DW, Woodward DF, Regan JW. Cloning of a carboxyl-terminal isoform of the prostanoid FP receptor. J Biol Chem 1997;272:883–7.

[19] Fujino H, Pierce KL, Srinivasan D, Protzman CE, Krauss AH-P, Woodward DF, et al. Delayed reversal of shape change in cells expressing FP_B prostanoid receptors. J Biol Chem 2000;275:29907–14.

[20] Vielhauer GA, Fujino H, Regan JW. Cloning and localization of h(FP)S: a six-transmembrane mRNA splice variant of the human FP prostanoid receptor. Arch Biochem Biophys 2004;421:175–85.

[21] Koda N, Tsutsui Y, Niwa H, Ito S, Woodward DF, Watanabe K. Synthesis of prostaglandin F ethanolamide by prostaglandin ethanolamide by prostaglandin F synthase and identification of bimatoprost as a potent inhibitor of the enzyme: new enzyme method using LC/ESI/MS. Arch Biochem Biophys 2004;424:128–36.

[22] Moriuchi H, Koda N, Okuda-Ashitaka E, Daiyasu H, Ogasawara K, Toh H, et al. Molecular characterization of a novel type of prostamide/prostaglandin F synthase, belonging to the thioredoxin-like superfamily. J Biol Chem 2008;283:792–801.

[23] Barski O, Tipparaju SM, Bhatnagar A. The aldo-keto reductase superfamily and its role in drug metabolism and detoxification. Drug Metab Rev 2008;40:553–624.

[24] Yang W, Ni J, Woodward DF, Tang-Liu DD-S, Ling K-HJ. Enzymatic formation of prostamide$F_{2\alpha}$ from anandamide involves a newly identified intermediate metabolite, prostamide H_2. J Lipid Res 2005;46:2745–51.

[25] Yoshikawa K, Takei S, Hasegawa-Ishii S, Chiba Y, Furukawa A, Kawamura N, et al. Preferential localization of prostamide/prostaglandin F synthase in myelin sheaths of the central nervous system. Brain Res 2011;1367:22–32.

[26] Gatta L, Piscitelli F, Giordano C, Boccella S, Lichtman A, Maione S, et al. Discovery of prostamide F2α and its role in inflammatory pain and dorsal horn nociceptive neuron hyperexcitability. PLoS ONE 2012;7:e31111.

[27] Khidir KG, Woodward DF, Farjo NP, Farjo BK, Tang ES, Wang JW, et al. The prostamide-related glaucoma therapy, bimatoprost, offers a novel approach for treating scalp alopecias. FASEB J 2013;27:557–67.

[28] Silvestri C, Martella A, Poloso NJ, Piscitelli F, Capasso R, Izzo A, et al. Anandamide-derived prostamideF2α negatively regulates adipogenesis. J Biol Chem 2013;288:23307–21.

[29] Woodward DF, Krauss AH, Nilsson SF. Bimatoprost effects on aqueous humor dynamics in monkeys. J Ophthalmol 2010;. 2010, ID 926192.

[30] Liang Y, Woodward DF., Human prostaglandin FP receptor variants and methods of using same. US patent 7320871, 2008.

[31] Duggan KC, Hermanson DJ, Musee J, Prusakiewicz JJ, Scheib JL, Carter BD, et al. (R)-profens are substrate-selective inhibitors of endocannabinoid oxygenation by COX-2. Nat Chem Biol 2011;7:803–9.

[32] Bishay P, Schmidt H, Marian C, Häussler A, Wijnvoord N, Ziebell S, et al. R-flurbiprofen reduces neuropathic pain in rodents by restoring endogenous cannabinoids. PLoS ONE 2010;5:e10628.

[33] Huggins JP, Smart TS, Langman S, Taylor L, Young T. An efficient, randomized placebo-controlled clinical trial with the irreversible fatty acid amide hydrolase-1 inhibitor PF-04457845, which modulates endocannabinoids but fails to induce effective analgesia in patients with pain due to osteoarthritis of the knee. Pain 2012;153:1837–46.

[34] Jakab B, Helyes Z, Varga A, Böleskei K, Szabó A, Sándor K, et al. Pharmacological characterization of the TRPVI receptor antagonist JYL1421 (SC0030) in vitro and in vivo in the rat. Eur J Pharmacol 2005;517:35–44.

[35] Moriyama T, Higashi T, Togashi K, Iida T, Segi E, Sugimoto Y, et al. Sensitization of TRPVI by EP_1 and IP reveals peripheral nociceptive mechanism of prostaglandins. Mol Pain 2005;1:3.

[36] Urquhart P, Wang J, Woodward DF, Nicolau A., Identification of prostamide E_2 fatty acyl ethanolamides and their biosynthesis precursors in rabbit cornea. J Lipid Res 2014, in press.

[37] Namura DK, Morrison BE, Blankman JL, Long JZ, Kinsey SG, Marcondes MCG, et al. Endocannabinoid hydrolysis generates brain prostaglandins that promote neuroinflammation. Science 2011;334:809–13.

[38] Taketani Y, Yamagishi R, Fujishiro T, Igarashi M, Sakata R, Aiara M., Activation of the prostanoid FP receptor inhibits adipogenesis leading to deepening of the upper eyelid sulcus in prostaglandin-associated periorbitopathy. Invest Ophthalmol Vis Sci 2014;55:1269–76.

[39] Peplinski LS, Albiani Smith K. Deepening of lid sulcus from topical bimatoprost therapy. Optom Vis Sci 2004;81:574–7.

[40] Filippopoulos T, Paula JS, Torun N, Hatton MP, Pasquale LR, Grosskreutz CL. Periorbital changes associated with topical bimatoprost. Ophthal Plast Reconstr Surg 2008;24:302–7.

[41] Yam JC, Yuen NS, Chan CW. Bilateral deepening of upper lid sulcus from topical bimatoprost therapy. J Ocul Pharmacol Ther 2009;25:471–2.

[42] Aydin S, Isikligil I, Teksen YA, Kir E. Recovery of orbital fat pad prolapsus and deepening of the lid sulcus from topical bimatoprost therapy: 2 case reports and review of the literature. Cutan Ocul Toxicol 2011;29:212–6.

[43] Choi HY, Lee JE, Lee JW, Park HJ, Jung JH. *In vitro* study of antiadipogenic profile of latanoprost, travoprost, bimatoprost, and tafluprost in human orbital preadiopocytes. J Ocul Pharmacol Ther 2011;28:146–52.

[44] Katz LJ, Cohen JS, Batoosingh AL, Felix C, Shu V, Schiffman RM. Twelve-month, randomized, controlled trial of bimatoprost 0.01%, 0.0125%, and 0.03% in patients with glaucoma or ocular hypertension. Am J Ophthalmol 2010;149:661–71.

[45] Craven ER, Liu C-C, Batoosingh AL, Schiffman RM, Whitcup SM. A randomized, controlled comparison of macroscopic conjunctival hyperemia in patients treated with bimatoprost 0.01% or vehicle who were previously controlled on latanoprost. Clin Ophthalmol 2010;4:1433–40.

[46] Pfennigsdorf S, Ramez O, von Kistowski G, Mäder B, Echstruth P, Fröböse M, et al. Multicenter, prospective, open-label, observational study of bimatoprost 0.01% in patients with primary open-angle glaucoma or ocular hypertension. Clin Ophthalmol 2012;6:739–46.

[47] Nixon DR, Simonyi S, Bhogal M, Sigouin CS, Crichton AC, Discepola M, et al. An observational study of bimatoprost 0.01% in treatment – naïve patients with primary open angle glaucoma or ocular hypertension: the CLEAR trial. Clin Ophthalmol 2012;6:2097–103.

[48] Chen J, Champa-Rodriguez ML, Woodward DF. Identification of a prostanoid FP receptor population producing endothelium-dependent vasorelaxation in the rabbit jugular vein. Br J Pharmacol 1995;116:3035–41.

[49] Chen J, Dinh T, Woodward DF, Holland JM, Yuan Y-D, Lin T-H, et al. Bimatoprost: mechanism of ocular surface hyperemia associated with topical therapy. Cardiovasc Drug Rev 2005;23:231–46.

[50] Paranhos A, Mendoça M, Silva MJ, Giampani J, Almeida Torres RJ, Della Paolera M, et al. Hyperemia reduction after administration of a fixed combination of bimatoprost and timolol maleate to patients on prostaglandin or prostamide therapy. J Ocular Pharmacol 2010;26:611–5.

[51] Katsanos A, Dastiridou AI, Fanariotis M, Kotoula M, Tsironi EE. Bimatoprost and bimatoprost/timolol fixed combination in patients with open-angle glaucoma and ocular hypertension. J Ocular Pharmacol 2011;27:67–71.

[52] Yamagishi R, Aihara M, Araie M. Neuroprotective effects of prostaglandin analogues on retinal ganglion cell death independent of intraocular pressure reduction. Exp Eye Res 2011;93:265–70.

[53] Takano N, Tsuruma K, Ohno Y, Shimazawa M, Hara H. Bimatoprost protects retinal neuronal damage via Akt pathway. Eur J Pharmacol 2013;702:56–61.

[54] Woodward DF, Tang ES-H, Attar M, Wang JW. The biodisposition and hypretricotic effects of bimatoprost in mouse skin. Exp Dermatol 2012;22:145–8.

CHAPTER 7

Prostamide $F_{2\alpha}$ Biosynthesizing Enzymes

Kikuko Watanabe, David F. Woodward

7.1 INTRODUCTION

The discovery that the mammalian endocannabinoids would be converted to a number of electrochemically neutral prostanoids introduced a new area of eicosanoid research. Although both prostaglandin (PG)-glyceryl esters and PG-ethanolamides (prostamides) are formed from 2-arachidonyl glycerol (2-AG) ester and anandamide [1,2], respectively, only prostamide $F_{2\alpha}$ has received consistent attention as a research topic. The prostamide $F_{2\alpha}$ analog bimatoprost, with its unique pharmacology and commercial prominence as an ocular hypotensive and hypertrichotic agent [3], has provided much of the research impetus. It follows that enzymes potentially involved in prostamide $F_{2\alpha}$ biosynthesis received attention; indeed, other PG synthases received no attention at all. The first enzymes demonstrated to be involved in prostamide $F_{2\alpha}$ biosynthesis

The Endocannabinoidome: The World of Endocannabinoids and Related Mediators. DOI: 10.1016/B978-0-12-420126-2.00007-9
Copyright © 2015 Elsevier Inc. All rights reserved

were the various aldo-keto reductases functioning as PGF synthases [4–11]. A second quite different enzyme, prostamide/PGF synthase, was discovered [20] comparatively recently and is gradually occupying center stage [3]. This review compares the properties and distribution of these two enzymes, predominately from the standpoint of $PGF_{2\alpha}$ and prostamide $F_{2\alpha}$. The known and putative pathways for the biosynthesis of prostamide $F_{2\alpha}$ and its stereoisomers are depicted in Figure 7.1.

7.2 PGF SYNTHASES BELONGING TO THE ALDO-KETO REDUCTASE SUPERFAMILY: MOLECULAR STRUCTURE AND PROPERTIES

The first PGF synthases to be fully characterized were the two isoforms found in lung and liver and their species homologs [4–11]. Following cloning of the human enzyme, it was designated AKR1C3 [11–13]. The mouse enzyme is designated AKR1C18: the rat enzyme AKR1C8 [11].

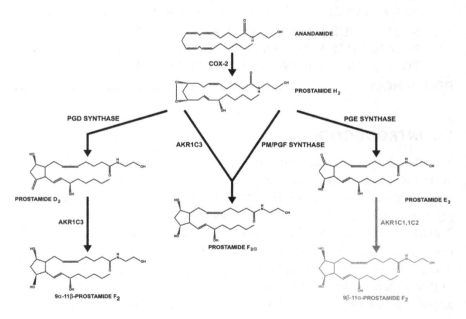

Fig. 7.1. Biosynthetic pathways to prostamide formation; putative pathways and products are depicted in red. 9α, 11α-Prostamide F_2 may be directly formed from prostamide H_2 by two enzymes, prostamide/PGF synthase [20] and AKR1C3 and related PGF synthases [4–14]. Prostamide D_2 may also be reduced by AKR1C3 and related PGF synthases to 9α, 11β-prostamide F_2. Prostamide E_2 may also be reduced by AKR1C1 and 1C2 but no evidence exists for this pathway to date.

These PGF synthases of the AKR superfamily are not substrate specific and various carbonyl compounds, such as phenanthrenequinone, may be reduced [4]. AKR1C3 and its congeners catalyze the conversion of PGH$_2$ to PGF$_{2\alpha}$ and reduce the 11-keto of PGD$_2$ to yield 9α, 11β-PGF$_2$ as a product [4–13]. The human recombinant PGF synthase was found to catalyze identical enzymatic reactions for the corresponding prostamides [10]. Thus, prostamide D$_2$ was converted to 9α, 11β-prostamide F$_2$ and a similar PGH$_2$ 9,11-endoperoxide reductase activity was apparent [10]. Evidence from inhibition studies suggests that these reactions are separate and occur at two different binding sites [10]. Bimatoprost was discovered as a potent inhibitor of all reactions [10] and was used to obtain the crystal structure of this PGF synthase [14]. The significance of PGF synthase inhibition by bimatoprost as an additional biological property will be addressed later in this review.

The cyclopentenone ring of PGE$_2$ is the mirror image of that of PGD$_2$; therefore, it is not unexpected that the 9-carbonyl group is similarly enzymatically reduced to produce 9β, 11α-PGF$_2$. This reaction is not subserved by PGF synthase [9,10] but rather this conversion requires a different aldo-keto reductase, again requiring NADPH as a cofactor [9]. This enzyme is designated PGE 9-keto reductase [9,15–19]. There is no evidence, to date, that PGE 9-keto reductase is involved in prostamide F$_2$ biosynthesis. The known and speculative aldo-keto reductase biosynthetic pathways to prostamide F$_2$ molecular species are compared in Figure 7.1. It has also been proposed the AKR1C1 and 1C2 may also catalyze the reduction of PGE$_2$ to PGF$_{2\alpha}$ [17].

7.3 PROSTAMIDE/PGF SYNTHASE: PROPERTIES

The realm of PGF$_{2\alpha}$ biosynthesis has been expanded and clarified by the comparatively recent discovery of a new enzyme, prostamide/PGF synthase [20]. This enzyme differs from the aldo-keto reductases in many regards. Prostamide/PGF synthase specifically recognizes PGH$_2$, while PGD$_2$ and PGE$_2$ are not substrates [20]. Equally, prostamide H$_2$ is a good substrate but broad specificity was nevertheless defined using t-butyl hydroperoxide and cumene hydroperoxide [20]. This enzyme is quite unique as a PGF$_{2\alpha}$/prostamide F$_{2\alpha}$ synthesizing enzyme in that reduced thioredoxin preferentially served as a reducing equivalent donor [20].

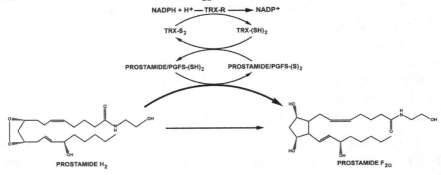

NADPH   NADP+

PROSTAMIDE H$_2$ PROSTAMIDE F$_{2\alpha}$

(b) PROSTAMIDE/PGF SYNTHASE: PROSTAMIDE F$_{2\alpha}$ BIOSYNTHESIS

NADPH + H+ — TRX-R ⟶ NADP+

TRX-S$_2$ TRX-(SH)$_2$

PROSTAMIDE/PGFS-(SH)$_2$ PROSTAMIDE/PGFS-(S)$_2$

PROSTAMIDE H$_2$ PROSTAMIDE F$_{2\alpha}$

Fig. 7.2. Schematic representation of the reductase mechanisms for (a) AKR1C3 and related PGF synthase; (b) prostamide/PGF synthase, (prostamide/PGFS). Cysteine residues (Cys^{44} and Cys^{47}) are oxidized to form a disulfide (prostamide/PGFS-S$_2$) and in an exchange reaction thioredoxin {Trx-(SH)$_2$} restores prostamide/PGFS. The oxidized disulfide form of thioredoxin (Trx-S$_2$) is then reduced by thioredoxin reductase (Trx-R), employing NADPH as a cofactor.

The reaction schematic comparing prostamide/PGF synthase with the aldo-keto reductase AKR1C3 is depicted in Figure 7.2. The crystal structure of prostamide/PGF synthase remains to be determined.

7.4 AKR1C3 AND RELATED PGF SYNTHASES AND PROSTAMIDE/PGF SYNTHASE: SUBSTRATES AND INHIBITORS

As previously mentioned, all of these reductases possess broad substrate specificity. In the context of prostanoid biochemistry, however, the specific recognition of PGH$_2$ and prostamide H$_2$ by prostamide/PGF synthase is of critical importance. This imparts a highly specific and direct biosynthetic pathway to PGF$_{2\alpha}$ and prostamide F$_{2\alpha}$.

Inhibitors of AKR1C3 are listed in Table 7.1 and activity is presented in terms of PGD$_2$, prostamide D$_2$, and PGH$_2$ reductase activities. Aldose reductase inhibitors were universally active against all reductase

Table 7.1 Inhibitor IC$_{50}$ Values (μM) on Prostamide F$_{2\alpha}$ Biosynthesizing Enzymes

Reductase Activities	Bimatoprost	ONO1373	ONO1349	ONO1370	Tolrestat	Stafil
PGD$_2$ 11-keto reductase activity	5	~1	~1	~1	100	12
Prostamide D$_2$ 11-keto reductase activity	60	~1	~1	~1	450	13
PGH$_2$ 9, 11-endoperoxide reductase activity	6	60	130	300	300	58

functions of AkRIC3 [10]. Bimatoprost was more potent but preferentially inhibited AkRIC3 reduction of PGH$_2$ and PGD$_2$ [10,14]. Reductase inhibition clearly requires a definitive structure–activity relationship, since analogs of PGH$_2$ selectively inhibit PGH$_2$ 9,11-endoperoxide reductase activity but do not alter PGD$_2$ on prostamide D$_2$ 11-keto reductase activity [10].

In contrast, no potent and selective inhibitor of prostamide/PGF synthase has been reported to date. The human aldose reductase AKR1B1 has also now been added to the repertoire of PGF synthases [21], but it is unknown whether any precursor prostamides behave as substrates.

7.5 COMPARATIVE DISTRIBUTION OF AKR1C3 AND RELATED AKR ENZYMES WITH PROSTAMIDE/PGF SYNTHASE

The biosynthetic products, PGF$_{2\alpha}$ and prostamide F$_{2\alpha}$, exhibit numerous biological activities but the distributions of prostamide/PGF synthase, AKR1C3 and related PGF synthases do not equate. In addition to biodistribution differences, the V_{max} for prostamide F$_{2\alpha}$ biosynthesis is five times greater for prostamide/PGF synthase compared to the aldo-keto reductase AKR1C3 [20]. This should be accounted for in consideration of regional expression of the two enzymes. Aldo-keto reductase PGF synthases of the AKR1C3 lineage are highly expressed in the lung and liver but in the brain and spinal cord exhibit low expression [4,8–12,14]. In contrast, prostamide/PGF synthase is most abundant in the brain and spinal cord [20]. Both enzymes are highly expressed in male and female reproductive organs [20]. The comparative expression in various animal tissues [6,7,20,22–24] is provided in Table 7.2. Further discussion on enzyme expression is provided in the next section and linked to function and therapeutics.

Table 7.2 Comparative Expression of Prostamide/PGF Synthase, AkRIC3 and Related PGF Synthases, and AKR1B1

Tissue	Prostamide/PGF Synthase	AkRIC3 and Related PGF Synthases	AkRIB1
Brain	✓✓✓		
Spinal cord	✓✓✓	✓✓	
Heart	✓✓✓	✓	
Thymus	✓✓✓		✓
Adrenal gland	✓✓		
Ovary	✓✓		
Uterus	✓✓	✓	✓
Testis	✓✓		
Vesicular gland	✓✓		✓
Vas deferens	✓		
Epididymidis	✓		
Lung	✓✓	✓✓✓	
Caecum	✓✓		
Large intestine	✓✓		
Small intestine	✓	✓	
Colon		✓	
Stomach	✓✓	✓	
Spleen	✓	✓✓	
Kidney	✓		
Eye	✓		
Liver	✓	✓✓✓	
Pancreas		✓	

Estimation based on enzymatic activities, Northern blot and Western blot analyses and immunocytochemistry [6,7,20–23].

7.6 PROSTAMIDE/PGF SYNTHASE AKR1C3 AND RELATED PGF SYNTHASES AND THEIR RELATION TO BIOLOGICAL FUNCTION AND THERAPEUTICS

Gene deletion technology has been extensively used to attribute physiological functions to the protein of interest. To date, prostamide/PGF synthase knock-out mice have not been reported. Gene deletion studies on species homologs of AKR1C3, 1C8, and 1C18, present their own difficulties. Interspecies nonuniformity imposes limitation on gene deletion technology, even in cases where the human homolog has been clearly identified in mice [11]. Close sequence homology between the different aldo-keto reductase also makes gene deletion problematical. The identification of these enzymes, therefore, relies on the combination of tissue distribution studies, detection of prostamide F$_{2\alpha}$ and its precursor anandamide, and functional studies on prostamide F$_{2\alpha}$ and its structural analog bimatoprost. This approach has provided the first unequivocal evidence for the importance of prostamide/PGF synthase and COX-2 in the biosynthesis of the spinal cord neurotransmitter prostamide F$_{2\alpha}$ [24,25].

The earliest studies intended to detect endogenous prostamide F$_{2\alpha}$ in tissues were guided by aldo-keto reductase PGF synthase expression, with results that were not entirely satisfactory [26]. The subsequent discovery of prostamide/PGF synthase and its high expression levels in the spinal cord [20] provided a signpost for endogenous prostamide F$_{2\alpha}$ discovery. Studies on the rat spinal cord revealed the presence of prostamide F$_{2\alpha}$, the levels of which were elevated in response to inflammation [25]. Prostamide F$_{2\alpha}$ was further established as a nociceptive neurotransmitter [25]. To date, this remains the singular case where prostamide/PGF synthase and its product prostamide F$_{2\alpha}$ have been established as functionally important at both the biochemical and pharmacological levels. Nevertheless, there are clues to the possible involvement of prostamide F$_{2\alpha}$ and its biosynthetic enzymes in other biological systems. In isolated human scalp hair follicles, careful inspection suggests that a prostamide antagonist attenuates hair shaft growth rate compared to control [27]. In adipogenesis, prostamide F$_{2\alpha}$ inhibits preadipocyte differentiation into adipocytes and provides a feed-forward loop, which opposes the proadipogenic effects of anandamide [28]. The feed-forward mechanism involves upregulation of COX-2 and prostamide receptors. Studies on 3T3-L1 cells showed AKR1B3, the mouse homolog of AKR1B1 [11] was decreased during the second day of differentiation [28]. Prostamide/PGF synthase expression remained unaltered but a role in adipogenesis cannot be excluded as COX-2 is the rate-limiting step in prostamide F$_{2\alpha}$ formation [28].

Other miscellany may portend a role for prostamide F$_{2\alpha}$ and its biosynthetic enzymes in regulating other cellular and tissue functions. Bimatoprost activity could potentially involve AKR1C3 inhibition, prostamide receptor activation, or both. The majority of bimatoprost effects reported so far appear to involve prostamide receptor involvement since the effects are mimicked by prostamide $F_{2\alpha}$ and/or the effects are blocked by prostamide receptor antagonists [3]. Bimatoprost and prostamide $F_{2\alpha}$ both contract certain smooth muscle preparations, decrease intraocular pressure, and attenuate cytokine damage to human colonic mucosal cells [3,29–32]. These findings merit further investigation, which would be greatly assisted by potent and selective enzyme inhibitors that lack the complicated pharmacology of bimatoprost.

REFERENCES

[1] Yu M, Ives D, Ramesha CS. Synthesis of prostaglandin E2 ethanolamide from anandamide by cyclooxygenase-2. J Biol Chem 1997;272:21181–6.

[2] Kozak KR, Crews BC, Morrow JD, Wang LH, Ma YH, Weinander R, et al. Metabolism of the endocannabinoids, 2-arachidonylglycerol and anandamide, into prostaglandin, thromboxane and prostacyclin glycerol esters and ethanolamides. J Biol Chem 2002;277: 44877–85.

[3] Woodward DF, Wang JW, Poloso NJ. Recent progress in prostaglandin F2α ethanolamide (prostamide F2α) research and therapeutics. Pharmacol Rev 2013;65:1135–47.

[4] Watanabe K, Yoshida R, Shimizu T, Hayaishi O. Enzymatic formation of prostaglandin $F_{2\alpha}$ from prostaglandin H_2 and D_2. J Biol Chem 1985;260:7035–41.

[5] Watanabe K, Iguchi Y, Iguchi S, Arai Y, Hayaishi O, Roberts LJ. Stereospecific conversion of prostaglandin D_2 to (5Z, 13E)-(15S)-9α-11β, 15-trihydrozyprosta-5,13-dien-1-oic acid(9α, 11β-prostaglandin F_2) and of prostaglandin H_2 to prostaglandin $F_{2\alpha}$ by bovine lung prostaglandin F synthase. Proc Natl Acad Sci USA 1986;83:1583–7.

[6] Hayashi H, Fujii Y, Wantanabe K, Urade Y, Hayaishi O. Enzymatic conversion of prostaglandin H_2 to prostaglandin $F_{2\alpha}$ by aldehyde reductase from human liver: comparison to the prostaglandin F synthetase from bovine lung. J Biol Chem 1989;264:1036–40.

[7] Urade Y, Watanabe K, Eguchi N, Fujii Y, Hayaishi O. 9α, 11β-prostaglandin $F_{2\alpha}$ formation in various bovine tissues. J Biol Chem 1990;265:12029–35.

[8] Suzuki T, Fujii Y, Miyano M, Chen L-Y, Takahashi T, Watanabe K. cDNA cloning, expression, and mutagenesis study of liver-type prostaglandin F synthase. J Biol Chem 1999;274:241–8.

[9] Watanabe K. Prostaglandin F synthase. PGs Other Lipid Med 2002;68–69:401–7.

[10] Koda N, Tsutsui Y, Niwa H, Ito S, Woodward DF, Watanabe K. Synthesis of prostaglandin F ethanolamide by prostaglandin F synthase and identification of bimatoprost as a potent inhibitor of the enzyme: new enzyme assay method using LC/ESI/MS. Arch Biochem Biophys 2004;424:128–36.

[11] Barski OA, Tipparaju SM, Bhatnagar A. The aldo-keto reductase superfamily and its role in drug metabolism and detoxification. Drug Metab Rev 2008;40:553–624.

[12] Matsuura K, Shiraishi H, Hara A, Sato K, Deyashiki Y, Ninomiya M, et al. Identification of a principal mRNA species for human 3 alpha-hydroxysteriod dehydrogenase isoform (AkRIC3) that exhibits high prostaglandin D-2 11-ketoreductase activity. J Biochem 1998;124: 940–6.

[13] Suzuki-Yamamoto T, Nishizawa M, Fukui M, Okuda-Ashitaka E, Nakajima T, Ito S, et al. cDNA cloning, expression and characterization of human prostaglandin F synthase. FEBS Lett 1999;462:335–40.

[14] Komoto J, Yamada T, Watanabe K, Woodward DF, Takusagawa F. Prostaglandin F$_{2\alpha}$ formation from prostaglandin H$_2$ by prostaglandin F synthase (PGFS): crystal structure of PGFS containing bimatoprost. Biochemistry 2006;45:1987–96.

[15] Wermuth B. Purification and properties of an NADPH-dependent carbonyl reductase from human brain. Relationship to prostaglandin 9-ketoreductase and xenobiotic ketone reductase. J Biol Chem 1981;256:1206–13.

[16] Schieber A, Frank RW, Ghisla S. Purification and properties of prostaglandin 9-ketoreductase from human kidney. Identity with human carbonyl reductase. Eur J Biochem 1992;206: 491–502.

[17] Wintergalen N, Thole HH, Galla HJ, Schlegel W. Prostaglandin-E2 9-reductase from corpus luteum of pseudopregnant rabbit is a member of the aldo-keto reductase superfamily featuring 20-hydroxsteroid dehydrogenase activity. Eur J Biochem 1995;234:264–70.

[18] Hayashi H, Fujii Y, Watanabe K, Hayaishi O. Enzymatic formation of prostaglandin-F2-alpha in human brain. Neurochem Res 1990;15:385–92.

[19] Dozier B, Watanabe K, Duffy D. Two pathways for prostaglandin F2 (alpha) (PGF2) (alpha) synthesis by the primate periovulatory follicle. Reproduction 2008;136:53–63.

[20] Moriuchi H, Koda N, Okuda-Ashitaka E, Daiyasu H, Ogasawara K, Toh H, et al. Molecular characterization of a novel type of prostamide/prostaglandin F synthase, belonging to the thioredoxin-like superfamily. J Biol Chem 2008;283:792–801.

[21] Bresson E, Lacroix-Pepin N, Boucher-Kovalik S, Chapdelaine P, Fortier MA. The prostaglandin F synthase activity of the human aldose reductase AkRIBI brings new lenses to look at pathologic conditions. Front Pharmacol 2012;3:1–13. article 98.

[22] Suzuki-Yamamoto T, Toida K, Tsurvo Y, Watanabe K, Ishimura K. Immunocytochemical localization of lung-type prostaglandin F synthase is the rat spinal cord. Brain Res 2002;877:391–5.

[23] Waclawik A, Rivero-Muller A, Blitek A, Kaczmarek MM, Brokken LJS, Watanabe K, et al. Molecular cloning and spatiotemporal expression of prostaglandin F synthase and microsomal prostaglandin E synthase-1 in porcine endometrium. Endocrinology 2006;47:210–21.

[24] Yoshikawa K, Takei S, Hasegawa-Ishii S, Chiba Y, Furukawa A, Kawamura N, et al. Preferential localization of prostamide/prostaglandin F synthase in myelin sheaths of the central nervous system. Brain Res 2011;1367:22–32.

[25] Gatta L, Piscitelli F, Giordano C, Boccella S, Lichtman A, Maione S, et al. Discovery of prostamide F2α and its role in inflammatory pain and dorsal horn nociceptive neuron hyperexcitability. PLoS ONE 2012;7:e31111.

[26] Weber A, Ni J, Ling K-HJ, Acheampong A, Tang-Liu DD-S, Cravatt BF, et al. Formation of prostaglandin 1-ethanolamides (prostamides) from anandamide in fatty acid amide hydrolase knockout (FAAH−/−) mice analyzed by high performance liquid chromatography with tandem mass spectrometry. J Lipid Res 2004;45:757–63.

[27] Khidir KG, Woodward DF, Farjo NP, Farjo BK, Tang ES, Wang JW, et al. The prostamide-related glaucoma therapy, bimatoprost, offers a novel approach for treating scalp alopecias. FASEB J 2013;27:557–67.

[28] Silvestri C, Martella A, Poloso NJ, Piscitelli F, Capasso R, Izzo A, et al. Anandamide-derived prostamide F2α negatively regulates adipogenesis. J Biol Chem 2013;288:23307–21.

[29] Woodward DF, Krauss AH-P, Wang JW, Protzman CE, Nieves AL, Liang Y, et al. Identification of an antagonist that selectively blocks the activity of prostamides (prostaglandinethanolamides) in the feline iris. Br J Pharmacol 2007;150:342–52.

[30] Matias I, Chen J, De Petrocellis L, Bisogno T, Ligresti A, Fezza F, et al. Prostaglandin eth-anolamides (prostamides): in vitro pharmacology and metabolism. J Pharmacol Exp Ther 2004;309:745–57.

[31] Burk RM, Woodward DF. Bimatoprost, a novel efficacious ocular hypotensive drug now recognized as a member of a new class of agents called prostamides. Drug Develop Res 2007;68:146–55.

[32] Nicotra LL, Vu M, Harvey BS, Smid SD. Prostaglandin ethanolamides attenuate damage in a human explant colitis model. PGs Other Lipid Med 2013;100–101:22–9.

CHAPTER 8

Metabolic Enzymes for Endocannabinoids and Endocannabinoid-Like Mediators

Natsuo Ueda, Kazuhito Tsuboi, Toru Uyama

8.1 INTRODUCTION

Endocannabinoids are a class of endogenous lipid molecules, which act as ligands for cannabinoid (CB) receptors [1]. Among several endocannabinoids so far reported, arachidonoylethanolamide (anandamide) [2] and 2-arachidonoylglycerol (2-AG) [3,4] are recognized to be two major ones, which have been well characterized in the past 20 years. Although both anandamide and 2-AG are arachidonic acid-containing molecules, they belong to fatty acid ethanolamides (*N*-acylethanolamines (NAEs)) or monoacylglycerols (MAGs), respectively. Anandamide functions as a partial agonist of CB receptor 1 and its endogenous levels are usually much lower than 2-AG, which acts as a full agonist of CB1 and CB2. Moreover, in animal tissues, anandamide is a quantitatively minor component among various NAEs, while 2-AG is a major MAG. These facts strongly suggest that 2-AG plays more important roles in the endocannabinoid signaling than anandamide [5]. Both endocannabinoids are formed from membrane glycerophospholipids, but their biosynthetic

The Endocannabinoidome: The World of Endocannabinoids and Related Mediators. DOI: 10.1016/B978-0-12-420126-2.00008-0
Copyright © 2015 Elsevier Inc. All rights reserved

pathways are totally different and are composed of different enzymes. Enzymes responsible for their degradations are also different. Thus, specific enzyme inhibitors are expected to separately control endogenous levels of anandamide and 2-AG. Anandamide appears to be formed and degraded together with other NAEs such as palmitoylethanolamide (PEA) and oleoylethanolamide (OEA). These saturated or monounsaturated NAEs do not act as endocannabinoids, but exhibit biological activities such as antiinflammatory, analgesic, and anorexic effects via different receptors (e.g., peroxisome proliferator-activated receptor (PPAR) α) [6,7]. In this chapter, we discuss recent findings on metabolic enzymes for NAEs and 2-AG in mammalian tissues.

8.2 ENZYMES FOR THE BIOSYNTHESIS OF N-ACYLETHANOLAMINES

The major biosynthetic pathway for NAEs is "the transacylation–phosphodiesterase pathway," which comprises two consecutive enzyme reactions [8,9] (Figure 8.1). The first reaction is the acylation of the amino group of ethanolamine phospholipids, including phosphatidylethanolamine (PE) and ethanolamine plasmalogen (plasmenylethanolamine). The resultant NAE phospholipids, represented by N-acyl-PE (NAPE), are a class of unique phospholipid molecules with three long hydrocarbon chains, showing some biological effects such as membrane stabilization, in addition to being NAE precursors [10]. Although the formation of N-arachidonoyl-PE at this step is essential in the anandamide biosynthesis, the enzyme, selectively transferring arachidonoyl chain, is not known. The second step in this pathway is the release of NAE by the hydrolysis of NAPE. This reaction may occur at one step by the catalysis of NAPE-hydrolyzing phospholipase D (NAPE-PLD) or through NAPE-PLD-independent multistep pathways in which multiple hydrolases are involved. Anandamide may also be formed by the condensation of free arachidonic acid with ethanolamine in the reverse reaction of fatty acid amide hydrolase (FAAH) or by nonenzymatic transfer of the arachidonoyl chain from arachidonoyl-CoA to ethanolamine [9].

N-Acyltransferase, which is detected in mammalian tissues such as brain and testis, is membrane-associated and stimulated by Ca^{2+} [11]. Its catalytically unique property is the utilization of glycerophospholipid rather than acyl-CoA as an acyl donor. The enzyme selectively abstracts an acyl chain from sn-1 position of the donor phospholipid, and

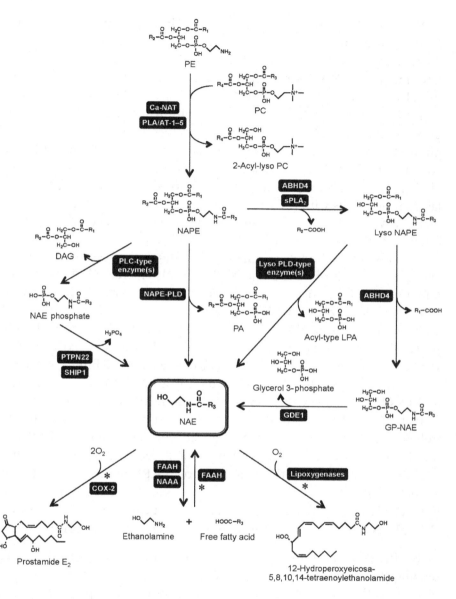

Fig. 8.1. Biosynthetic pathways of NAEs from PE and their degradative pathways., routes which are reported for anandamide. Ca-NAT, Ca²⁺-dependent N-acyltransferase.*

transfers this chain to the amino group of PE. It is suggested that the increase in intracellular Ca^{2+} concentration enhances the activity, leading to the accumulation of NAPE and NAE. However, Ca^{2+}-dependent *N*-acyltransferase has not been highly purified or cloned.

Since lecithin retinol acyltransferase (LRAT) is another acyltransferase abstracting an acyl group from sn-1 position of lecithin (phosphatidylcholine (PC)), it was likely that N-acyltransferase is structurally similar to LRAT. Our recent studies revealed that all members of the HRAS-like suppressor (HRASLS) family, which show homology with LRAT, have a NAPE-forming N-acyltransferase activity as well as phospholipase (PL)A_1/A_2 activity [9]. Therefore, we proposed to call this family the phospholipase A/acyltransferase (PLA/AT) family [12]. We constructed COS-7 cells transiently expressing one of PLA/AT-1–5 or HEK293 cells stably expressing PLA/AT-1 or -2. When we metabolically labeled these cells with [^{14}C]ethanolamine, we found that the expression of PLA/AT-1, -2, -4, and -5 significantly increases intracellular levels of [^{14}C]NAPE [13,14]. When HEK293 cells stably expressing PLA/AT-1 or -2 were analyzed by liquid chromatography–tandem mass spectrometry, endogenous levels of NAPE and NAE were also increased. ATDC5 cells and HeLa cells endogenously express PLA/AT-1 and -2, respectively. The knockdown of PLA/AT-1 or -2 in these cells by RNAi decreased endogenous levels of NAPE. These results suggest that PLA/AT family proteins at least in part contribute to the production of NAPE *in vivo*. Human tissues express all five members of the PLA/AT family, while rodents express only PLA/AT-1, -3, and -5. It is noted that humans, mice, and rats highly express PLA/AT-1 in brain, heart, skeletal muscle, and testis [12]. NAPE levels in these tissues are reported to increase during ischemia and inflammation. However, the enzyme activity of PLA/AT family proteins is Ca^{2+}-independent, and it remains unclear whether PLA/AT-1 is involved in the generation of NAPE upon cellular stimuli in these tissues. Interestingly, the overexpression of PLA/AT-2 and -3, but not -1, in mammalian cells causes disappearance of peroxisomes [13–15]. Its molecular mechanism, however, remains to be solved.

NAPE-PLD is a membrane-associated protein belonging to the metallo-β-lactamase family [16]. So far, this is the sole enzyme to release NAE directly from NAPE in mammalian tissues. The presence of N-acyl moiety is indispensable as substrates for NAPE-PLD since major membrane glycerophospholipids such as PE and PC are inactive [17]. In contrast, NAPE-PLD does not distinguish the length of N-acyl species of NAPE. NAPE-PLD thus converts various NAPE species with different N-acyl species to the corresponding NAE species at similar rates. NAPE accumulates in the brain of NAPE-PLD-deficient mice, showing the central role of NAPE-PLD in the metabolism of NAPE in this tissue [18,19]. However, any abnormality in phenotype has not been reported. The deficiency of NAPE-PLD caused a decrease in brain levels of saturated NAEs, but their total suppression did not occur. Furthermore, those of polyunsaturated NAEs, including anandamide, were unaltered [18]. This result demonstrated the presence of alternative pathway(s) for NAE formation *in vivo*. Another possible role of NAPE-PLD is the degradation of potentially harmful aldehyde-modified PEs such as levuglandin-modified PE [20]. Such modification of PE occurs nonenzymatically by oxidative stress, and the products are structurally similar to NAPE. The knockdown of NAPE-PLD expression increased the intracellular levels of levuglandin-modified PE, and recombinant NAPE-PLD hydrolyzed this modified PE to levuglandin-ethanolamine. Regarding the regulation of activity, NAPE-PLD is stimulated by a millimolar order of divalent cations, including Ca^{2+}, *in vitro* [21]. However, there is no evidence that NAPE-PLD activity is physiologically regulated by intracellular Ca^{2+}. As for transcriptional regulation, lipopolysaccharide downregulates NAPE-PLD expression in macrophages by altering the acetylation state of histone bound to the promoter of NAPE-PLD gene and the transcription factor Sp1 is involved in the regulation of the baseline expression of NAPE-PLD [22]. In a Norwegian population-based cohort study, a common haplotype in NAPE-PLD was shown to be protective against severe obesity [23].

So far, three pathways have been reported as NAPE-PLD-independent multistep pathways, which form NAE from NAPE (Figure 8.1) [9]. A route via *N*-acyl-lysophosphatidylethanolamine (lyso NAPE) and glycerophospho (GP)-NAE was proposed in 1980s [24]. We later showed that group IB, IIA, and V of the secretory PLA_2 ($sPLA_2$) isoforms produce 1-acyl-lyso NAPE from NAPE and that NAE can be released from lyso NAPE by a membrane-bound lyso PLD-type enzyme(s) [25]. Furthermore, the double *O*-deacylation of NAPE to GP-NAE via lyso NAPE and further hydrolysis of GP-NAE to NAE were proposed [26]. In fact, GP-NAE was detected in mouse brain [27] and NAPE-PLD-deficient mice showed a remarkable increase in brain levels of lyso NAPE and GP-NAE [19]. α/β-Hydrolase 4 (α/β-hydrolase domain-containing protein 4 (ABHD4)), a serine hydrolase, was shown to catalyze the double

O-deacylation of NAPE [26]. Namely, ABHD4 hydrolyzes the ester bonds of both NAPE and lyso NAPE. Notably, as a lysophospholipase substrate, ABHD4 prefers lyso NAPE to other lysophospholipids such as lyso PE, lyso PC, and lysophosphatidylserine (LPS). Recently, knockdown of ABHD4 in prostate epithelial cells was reported to inhibit anoikis (cell death in response to loss of cell–cell and cell–matrix interactions) [28]. It is unclear whether this effect is caused by altered NAE metabolism.

GDE1, a member of the glycerophosphodiester phosphodiesterase (GDE) family, is an integral membrane glycoprotein, which was originally reported as MIR16, a protein interacting with RGS16 (a regulator of G protein signaling) [29]. Later, GDE1 was shown to have a Mg^{2+}-dependent phosphodiesterase activity selectively hydrolyzing glycerophosphoinositol [30]. GDE1 also catalyzes the hydrolysis of GP-NAE to NAE [27]. Thus, a combination of ABHD4 and GDE1 forms a NAPE-PLD-independent pathway in mammalian tissues, including brain, which express both enzymes. However, it is unlikely that this pathway selectively generates anandamide, since neither ABHD4 nor GDE1 showed preference to precursors of anandamide. It should be noted, according to our analysis of NAPE-PLD-deficient mice [19], the brain level of anandamide was significantly decreased in contrast to the aforementioned result [18]. GDE1-deficient mice were born at the expected Mendelian frequency, were viable and healthy, and showed no abnormal cage behavior [31]. Although the brain homogenates hardly converted GP-NAE and lyso NAPE to NAE, endogenous brain levels of NAEs were not significantly decreased. Moreover, double knockout mice of GDE1 and NAPE-PLD did not exhibit a decrease in NAE levels, suggesting that additional enzymes or pathways are involved in the NAE formation. Interestingly, in GDE1-deficient mice, brain levels of glycerophosphoinositol, glycerophosphoserine, and glycerophosphoglycerate were remarkably increased [32]. A concomitant decrease in free serine levels was observed, suggesting the physiological importance of the glycerophosphoserine hydrolysis by GDE1 in serine homeostasis.

In rat brain, 65% of N-arachidonoylethanolamine phospholipids as anandamide precursors were plasmalogen-type (N-arachidonoyl-plasmenylethanolamine) [33]. The hydrolysis of N-palmitoyl-plasmenylethanolamine to PEA by recombinant NAPE-PLD proceeded at 70% of the rate of N-palmitoyl-PE hydrolysis [19]. Although N-acyl-plasmenylethanolamines and their lyso forms accumulated in the brain of **NAPE-PLD**-deficient mice, the brain homogenate of these mice could convert *N*-palmitoyl-plasmenylethanolamine to **PEA** via *N*-palmitoyl-lysoplasmenylethanolamine (Figure 8.2). This route requires a lyso PLD-type hydrolase, which

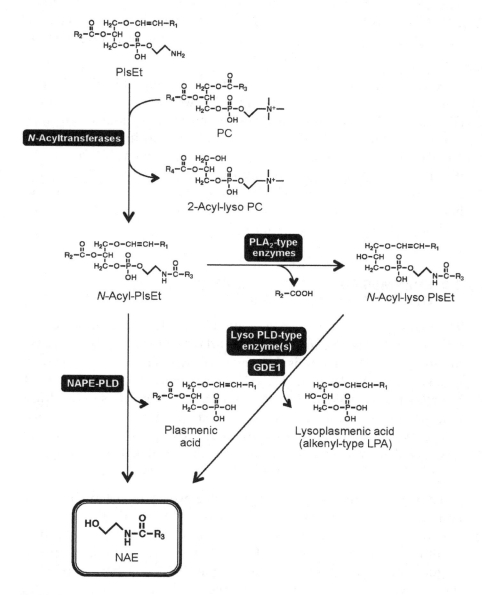

Fig. 8.2. Biosynthetic pathways of NAEs from plasmenylethanolamine (PlsEt).

directly releases PEA from N-palmitoyl-lysoplasmenylethanolamine, since the vinyl ether bond at the sn-1 position cannot be hydrolyzed by lysophospholipases such as ABHD4. We found that recombinant GDE1 exhibits this lyso PLD activity, although the activity is much lower than the aforementioned GP-NAE-hydrolyzing activity [19]. In brain, GDE1 appears to be at least in part responsible for lyso PLD activity, which produces PEA, OEA, and anandamide from their corresponding N-acyl-lysoplasmenylethanolamines.

Another multistep route to generate NAE from NAPE is the formation of NAE phosphate by PLC, followed by dephosphorylation (Figure 8.1). PTPN22 (protein tyrosine phosphatase, nonreceptor type 22) [34] and Src homology 2 domain-containing inositol-5-phosphatase 1 (SHIP1) [35] were identified as phosphatases responsible for the dephosphorylation of anandamide phosphate, while the PLC-type enzyme remains unidentified. This pathway was suggested to function in lipopolysaccharide-treated RAW264.7 mouse macrophages [34] and the brain of NAPE-PLD-deficient mice [35].

8.3 ENZYMES FOR THE DEGRADATION OF N-ACYLETHANOLAMINES

The major pathway for the degradation of NAEs is hydrolysis to free fatty acids and ethanolamine (Figure 8.1). The central role of FAAH in this hydrolysis reaction is demonstrated by a remarkable increase in endogenous NAE levels in FAAH-deficient mice and specific FAAH inhibitor-treated animals [36–38]. Similar to free polyunsaturated fatty acids, polyunsaturated NAEs such as anandamide can be converted to hydroxy or hydroperoxy derivatives or prostaglandin-like molecules (prostamides) by oxygenases such as cytochrome P-450 [39], lipoxygenases [40,41], and cyclooxygenase-2 [42,43], respectively.

FAAH is a membrane-associated serine hydrolase, belonging to the amidase signature family [44]. FAAH is widely distributed in mammalian tissues, and its expression level in rats is the highest in liver, small intestine, and brain [45]. FAAH is also called FAAH-1 to distinguish between FAAH and its isozyme FAAH-2 [46]. Ser-241 is the catalytic nucleophile and forms the catalytic triad together with

Lys-142 and Ser-217 [47]. Members of the amidase signature family, including FAAH, share a common core fold, which is comprised of a twisted β-sheet consisting of 11 mixed strands surrounded by a number of α-helices [48]. FAAH was crystallized as a dimer, and a helix-turn-helix motif (α18 and α19) appeared to be the hydrophobic membrane-binding domain [49]. Although anandamide is the most active substrate, FAAH also hydrolyzes other long-chain NAEs such as PEA and OEA [50]. Moreover, different classes of bioactive fatty acid amides, including oleamide (primary amide of oleic acid) [44] and *N*-acyltaurine (taurine-conjugated fatty acid) [51], serve as FAAH substrates. Such a broad substrate specificity of FAAH shows a possibility that some of the symptoms observed in FAAH-deficient or specific FAAH inhibitor-treated mice are caused by an increase in endogenous levels of CB receptor-insensitive bioactive fatty acid amides. In spite of a high esterase activity of recombinant FAAH toward 2-AG [52], its contribution as 2-AG hydrolase is minor in brain [53]. Among naturally occurring single nucleotide polymorphisms in human FAAH gene, the cytosine 385 to adenine missense mutation, resulting in the generation of P129T variant, was strongly associated with street drug use and problem drug/alcohol use [54]. The P129T variant exhibited an enhanced sensitivity to proteolytic degradation, suggesting that the hypoactivity of FAAH is linked to drug abuse and dependence. A catalytically inactive, truncated variant of FAAH was recently reported to have the ability to bind to anandamide and to drive anandamide transport in neurons and was named FAAH-like anandamide transporter (FLAT) [55]. However, this finding was questioned for certain reasons, including the lack of endogenous FLAT expression in mouse tissues and the detection of residual anandamide-hydrolyzing activity in recombinant FLAT [56]. Free fatty acids in tissue extracts inhibit FAAH activity [45]. Recently, monounsaturated fatty acids generated via stearoyl-CoA desaturase-1, but not derived from diets, were reported to be endogenous FAAH inhibitors mediating the high fat diet-induced increase in hepatic anandamide [57].

FAAH-2 shows ~20% sequence identity with FAAH-1 at amino acid level, and is present in human, but not in rodents [46]. FAAH-1 was localized on the outer face of mitochondria and endoplasmic reticulum [58], while FAAH-2 was localized on lipid droplets [59]. Its N-terminal hydrophobic region was shown to be a lipid droplet localization sequence. These results suggest a unique function of FAAH-2, which is distinct from that of FAAH-1.

Many specific FAAH inhibitors with therapeutic potential have been developed [38], which are represented by URB597 [60], OL-135 [61], PF-3845 [62], and PF-04457845 [63]. FAAH inhibitors are expected to show beneficial effects on a variety of diseases such as pain, inflammation, and sleeping disorders. Moreover, the administration to mice of organophosphorus nerve agents as dual inhibitors of FAAH and MAG lipase (MAGL) showed more than 10-fold increases in brain levels of both anandamide and 2-AG, and CB1-dependent behavioral effects called "tetrad of cannabinoid" (analgesia, hypomotility, hypothermia, and catalepsy) [64]. This full spectrum of cannabimimetic activities was not observed upon inhibition of either FAAH or MAGL. JZL195 as a selective and efficacious dual FAAH/MAGL inhibitor also mimicked the pharmacological activities of CB1 receptor agonist *in vivo* [65].

We previously found an anandamide-hydrolyzing activity in CMK human megakaryoblastic cells, which was characterized by its acidic pH optimum [66]. A similar activity was detected in rat lung and other tissues [67]. Protein purification, cDNA cloning, and morphological analysis revealed that this enzyme, named NAE-hydrolyzing acid amidase (NAAA), is a lysosomal glycoprotein belonging to the N-terminal nucleophile (Ntn) hydrolase superfamily [68–70]. Human NAAA has four *N*-glycosylation sites [71,72]. This protein has been recognized as acid ceramidase-like protein because of its high sequence homology with acid ceramidase, a lysosomal enzyme hydrolyzing ceramide to sphingosine and fatty acid [73]. On the other hand, there is no homology between NAAA and FAAH. The optimal pH is 4.5–5 and the activity is low or undetectable at neutral and alkaline pH, indicating that NAAA functions mainly inside lysosomes. When overexpressed in mammalian cells, recombinant human NAAA is first produced as an inactive proenzyme and is then matured to an active heterodimer by autocatalytic cleavage between Phe-125 and Cys-126. The N-terminal 28 amino acids, forming a signal peptide, are removed during maturation [72]. Cys-126 is considered to be the catalytic nucleophile and required for both NAE hydrolysis and autocatalytic cleavage. This cysteine residue is located at the N-terminus of the second subunit of heterodimer and is acylated by β-lactones acting as NAAA inhibitors [74,75].

NAAA hydrolyzes various NAEs with a preference for PEA. In addition, NAAA shows a relatively low ceramide-hydrolyzing activity. Notably, acid ceramidase also hydrolyzes NAE as well as ceramide [68]. Triton X-100 or Nonidet P-40 has been used to enhance NAAA activity. These nonionic detergents could be replaced with choline- or ethanolamine-containing phospholipids (PC, PE, and sphingomyelin) [76]. Similarly, the thiol reducing agent dithiothreitol, used as another stimulator, could be replaced with dihydrolipoic acid (the reduced form of α-lipoic acid). These endogenous molecules may be useful to maintain the active state of NAAA in lysosomes. NAAA is detected in various tissues of human and rodents with higher expression levels in macrophages [77,78] and prostate [79]. A possible compensatory induction of NAAA in the absence of FAAH was ruled out because the expression levels of NAAA mRNA in various mouse tissues were not significantly elevated in FAAH-deficient mice [76].

Since NAAA preferentially hydrolyzes PEA over anandamide, selective NAAA inhibitors may increase local levels of endogenous PEA, showing antiinflammatory and analgesic effects, without affecting the endocannabinoid system [80]. Among the reported NAAA inhibitors [9], lactone derivatives appear to be superior in potency and selectivity to others. These compounds include (S)-N-(2-oxo-3-oxetanyl)-3-phenylpropionamide [(S)-OOPP], (S)-N-(2-oxo-3-oxetanyl)biphenyl-4-carboxamide, (2S,3R)-2-methyl-4-oxo-3-oxetanylcarbamic acid 5-phenylpentyl ester (URB913/ARN077), and (4-phenylphenyl)-methyl-N-[(2S,3R)-2-methyl-4-oxo-oxetan-3-yl]carbamate (IC_{50} = 420, 115, 127, and 7 nM, respectively) [81–84]. (S)-OOPP increased PEA levels in activated leukocytes and macrophage cells, reduced neutrophil migration, and inhibited carrageenan-induced plasma extravasation [81]. Furthermore, when heat hyperalgesia and mechanical allodynia were elicited in mice by carrageenan injection or sciatic nerve ligation, topical administration of ARN077 attenuated these nociceptive responses through PPARα [85]. Although the physiological significance of NAAA remains unclear, these results suggest its involvement in the regulation of local PEA levels. However, when developing NAAA inhibitors, it should be taken into consideration that acid ceramidase is similar to NAAA in structure and catalytic mechanism.

8.4 ENZYMES FOR THE BIOSYNTHESIS OF 2-AG

Among the proposed 2-AG-biosynthesizing pathways, the phosphatidylinositol (PI)-PLC/diacylglycerol lipase (DAGL) pathway is considered to be predominant at least in the central nervous system (Figure 8.3) [86–88]. This pathway comprises two hydrolysis reactions. First, arachidonic acid-containing PI 4,5-bisphosphate is hydrolyzed

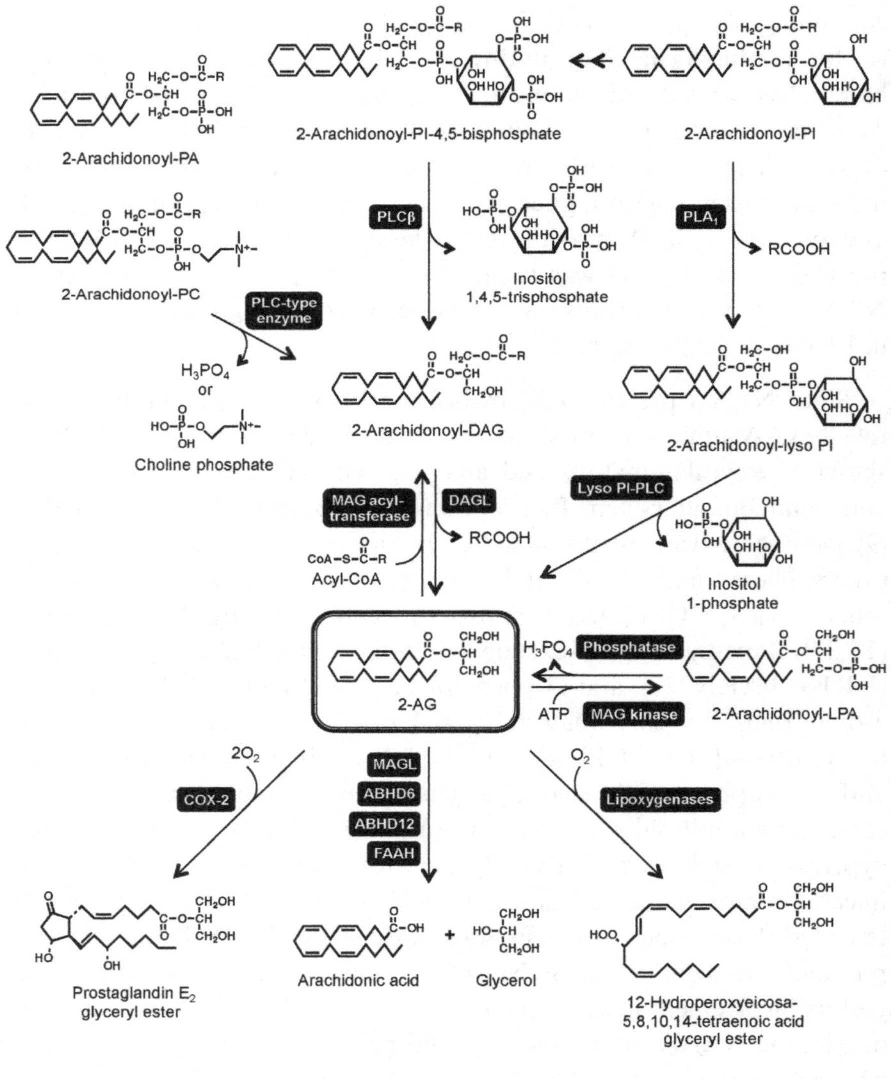

Fig. 8.3. Biosynthetic and degradative pathways of 2-AG.

to 1,2-diacylglycerol (DAG) and inositol 1,4,5-trisphosphate by PLC. Second, 2-arachidonoyl-DAG is hydrolyzed to 2-AG and free fatty acid by DAGL. Thirteen PLC isozymes identified are categorized into six classes, the β (1–4), γ (1, 2), δ (1, 3, 4), ϵ, ζ, and η (1, 2) types [89]. The β-type isozymes are characterized by hormonal stimulation via α subunit of $G_{q/11}$ protein. Among various $G_{q/11}$-coupled receptors in brain, PLCβ1 was coupled to the group I metabotropic glutamate receptor mGluR5 and the muscarinic acetylcholine receptor M1, while β3 and β4 were coupled to mGluR1 [87]. The analyses of PLCβ-deficient mice revealed that β1 and β4 isozymes are responsible for the G protein-coupled receptor-dependent 2-AG generation in hippocampal neurons and cerebellar Purkinje cells, respectively [90,91]. In addition to PI 4,5-bisphosphate, phosphatidic acid (PA) [92] and PC [93] were suggested to be precursors of 2-arachidonoyl-DAG (Figure 8.3). Moreover, 2-AG may be produced by the following pathways bypassing DAG: 1) hydrolysis of PI by PLA$_1$, followed by hydrolysis of the resultant lyso PI to 2-AG by lyso PI-specific PLC [94]; and 2) dephosphorylation of arachidonic acid-containing lysophosphatidic acid (LPA) to 2-AG, which is observed in rat brain homogenate [95] (Figure 8.3).

cDNA cloning of human DAGL by homology screening with *Penicillium* DAGL clarified that this enzyme is transcribed from two closely related genes α and β [96,97]. The deduced primary structures consist of 1042 and 672 amino acids, respectively, and α isoform is longer due to the presence of a long C-terminal tail. These structures possess a lipase-3 motif, a serine lipase motif, and four putative transmembrane domains. Both the α and β isoforms release a free fatty acid from *sn*-1 rather than *sn*-2 position of DAG to form 2-MAG selectively and are ubiquitously expressed in animal tissues. Their distribution in neurons largely shifts from elongating axons in the embryo to dendrites and dendritic spines after birth, suggesting that the roles of 2-AG signaling undergo a change depending on development [96,97]. $G_{q/11}$-coupled receptors, α subunit of $G_{q/11}$ protein, PLCβ, and DAGLα are colocalized at particular synaptic and neuronal surface [87]. DAGLα was involved in the mGluR-dependent 2-AG mobilization in neuroblastoma cells [98]. In this study DAGLα was found to interact with Homer-1b and Homer-2, two components of the molecular scaffold that enables the group I mGluR signaling.

Both DAGLα-deficient and DAGLβ-deficient mice were viable and their general appearances were normal [99,100]. In DAGLα-deficient mice, the basal brain levels of 2-AG were remarkably reduced and the stimulus-induced 2-AG elevation was not observed [99]. As expected, Ca^{2+}-dependent or $G_{q/11}$-coupled receptor-driven retrograde synaptic suppression was absent in the cerebellum, hippocampus, and striatum. By contrast, DAGLβ-deficient mice showed normal brain levels of 2-AG and intact retrograde synaptic suppression. In another set of DAGL-deficient mice, the deficiency of DAGLα or β reduced brain 2-AG levels by ~80% or by 50%, respectively [100]. The 2-AG formed by DAGLα was strongly suggested to be responsible for neurogenesis as well as retrograde suppression at central synapses. Recently, it was reported that DAGLβ also participates in 2-AG production in autaptic neurons [101]. In contrast to brain, DAGLβ deficiency decreased 2-AG levels in liver by 90%. Exposure of mice to ethanol diet upregulated DAGLβ selectively in hepatic stellate cells [102]. As for regulatory mechanism, it was reported that the activated Ca^{2+}/calmodulin-dependent protein kinase II (CaMKII) inhibits DAGLα activity by phosphorylating two serine residues in the C-terminal tail of DAGLα [103]. In fact, genetic inhibition of CaMKII increased striatal DAGL activity and basal 2-AG levels *in vivo*.

Although 1,6-bis(cyclohexyloximinocarbonylamino)hexane (RHC80267) and tetrahydrolipstatin are nonspecific DAGL inhibitors, their pharmacological effects shared by both compounds were shown to be good evidence for DAGL-dependent events [104]. O-3841 and O-5596 were reported to be more potent and selective DAGL inhibitors [105,106].

8.5 ENZYMES FOR THE DEGRADATION OF 2-AG

Several pathways have been reported for the degradation of 2-AG (Figure 8.3). The most important pathway is its hydrolysis to free arachidonic acid and glycerol by several hydrolases including MAGL. In addition, 2-AG can be phosphorylated to LPA by MAG kinase [107] or acylated to DAG by MAG acyltransferase using acyl-CoA as acyl donor [108]. Furthermore, similar to anandamide, the arachidonoyl moiety of 2-AG can be oxygenated to glyceryl prostaglandins by COX-2 or to hydroperoxy derivatives by lipoxygenases [109,110]. It should be noted that 2-AG can be directly converted to other bioactive lipids such as LPA and glyceryl prostaglandins [88].

MAGL is a serine hydrolase belonging to the α/β-hydrolase super-family with a catalytic triad (Ser-122, in a GXSXG motif, Asp-239, and His-269) and hydrolyzes both 1- and 2-MAGs, including 2-AG [111]. Its cDNA was cloned from mouse [112], human [113], and rat [114]. MAGL is a soluble protein, which binds to membranes in a peripheral manner [111], and is crystallized as a dimer [115]. In the tertiary structure, the catalytic triad, an apolar helix covering the active site, a wide and hydrophobic pocket for acyl binding, and a hydrophilic cavity for the glycerol moiety of 2-AG were identified. MAGL is ubiquitously expressed in mouse tissues [112] and its expression levels in rat brain are high in cerebellum, cortex, and hippocampus; medium in thalamus and striatum; and low in brain stem and hypothalamus [114]. MAGL plays the central role in the degradation of 2-AG responsible for endocannabinoid signaling in brain, which is demonstrated by many experimental results, including immunodepletion of 2-AG-hydrolyzing activity, immunohistochemistry exhibiting the presynaptic localization of MAGL, overexpression or silencing of MAGL at cell levels, and genetic or pharmacological deficiency of MAGL at whole-body level [114,116–118]. A functional proteomic approach showed that MAGL is responsible for ~85% of the 2-AG-hydrolyzing activity at pH 7.5 in mouse brain [53]. The remaining activity was mostly attributed to ABHD6 and ABHD12. The 2-AG-hydrolyzing activity in the brain of MAGL-deficient mice decreased by ~90% and its endogenous 2-AG level increased about 10-fold [117]. Similar results were observed in various peripheral tissues such as paw skin, spleen, heart, kidney, and liver of the mice. Importantly, repeated administration of the MAGL inhibitor JZL184 to mice caused the analgesic tolerance to CB1 receptor agonists, which was mimicked by genetic MAGL disruption. Chronic MAGL blockade also resulted in physical dependence, impairment of endocannabinoid-dependent synaptic plasticity, and desensitization of brain CB1 receptors [117]. However, such tolerance was not observed by prolonged blockage of FAAH, indicating the importance of a combination of 2-AG and MAGL in the CB1-mediated endocannabinoid system.

In addition to being an endocannabinoid, 2-AG, together with other MAGs, is well known as an intermediate in the metabolism of triacylglycerols and glycerophospholipids. Thus, the hydrolysis of MAGs by MAGL is involved in the mobilization of stored triacylglycerol in adipocytes as well as the release of arachidonic acid from membrane phospholipids [112,119]. In fact, MAGL deficiency impaired lipolysis, attenuated diet-induced insulin resistance [120], and decreased free arachidonic acid levels [121]. The arachidonic acid produced by MAGL was further converted to inflammatory prostaglandins in brain [122]. Moreover, high expression of MAGL resulted in malignant progression of cancer cells [123].

Since MAGL inhibitors have therapeutic potential, for example, as analgesics and anticancer agents, various serine hydrolase inhibitors and sulfhydryl blockers have been reported to inhibit MAGL nonspecifically or specifically [124–126]. The potent and selective MAGL inhibitors so far reported include URB602 [127], CAY10499 [128], OMDM169 [129], JZL184 [121], and KML29 [130]. Although repeated high dose administration of JZL184 caused tolerance, repeated low doses reduced inflammatory nociception through both CB1 and CB2 without apparent tolerance [131]. JZL195 as a selective and efficacious dual FAAH/MAGL inhibitor was mentioned above.

ABHD6 and ABHD12 are two integral membrane proteins belonging to the α/β-hydrolase superfamily [132,133]. ABHD6 received attention because of its expression in BV-2 mouse microglial cells, which lack MAGL [134]. In adult mouse cortex, ABHD6 was mostly located postsynaptically, in contrast to presynaptic localization of MAGL. The proposed role of ABHD6 as a 2-AG hydrolase was to limit the intracellular accumulation of 2-AG. UCM710 increased both anandamide and 2-AG levels in neurons by inhibiting FAAH and ABHD6 but not MAGL [135]. Recently, it was reported that knockdown of ABHD6 in peripheral tissues protects mice from high fat diet-induced obesity, hepatic steatosis, and systemic insulin resistance [136]. In addition to MAG-hydrolyzing activity, ABHD6 showed a lysophospholipase activity, exhibiting a preference for lysophosphatidylglycerol (LPG). Inhibition of ABHD6 led to the accumulation of LPG and phosphatidylglycerol *in vivo*. On the other hand, mutations in the ABHD12 gene were found to cause a neurodegenerative disease termed PHARC (polyneuropathy, hearing loss, ataxia, retinitis pigmentosa, and cataract) [137]. ABHD12-deficient mice exhibited massive increases in very long chain LPS species and ABHD12 functioned as a principal lysophospholipase for LPS in the mammalian brain [138]. In ABHD12-deficient mice, the brain LPS levels were elevated in early life, which was followed by age-dependent increases in microglial activation as well as auditory and motor defects similar to the symptoms of human PHARC patients. These results strongly suggest physiological roles of ABHD6 and ABHD12 as lysophospholipases apart from their 2-AG hydrolase activities.

8.6 PERSPECTIVES

In this chapter, we briefly introduced enzymes responsible for the metabolism of anandamide and 2-AG in mammalian tissues. In addition to well-known enzymes such as N-acyltransferase, NAPE-PLD, FAAH, PLCβ, DAGL, and MAGL, recent studies revealed the involvement of several additional enzymes such as ABHD4, ABHD6, ABHD12, GDE1, NAAA, and PLA/AT family members. However, since the substrates of these enzymes are not necessarily limited to endocannabinoids or their precursors and endocannabinoids have roles other than CB receptor ligands (e.g., ligands for other receptors or intermediates in lipid metabolism), we should be cautious when interpreting the molecular mechanisms of phenotypes, which are caused by genetic or pharmacological disruption of a particular enzyme. Currently, endocannabinoid-hydrolyzing enzymes such as FAAH, MAGL, and NAAA attract much attention as promising therapeutic targets. The physiological roles of NAEs, other than anandamide and their phospholipid precursors, are also interesting subjects to be explored. Thus, more thorough studies on the endocannabinoid-related enzymes will be required.

REFERENCES

[1] Pacher P, Bátkai S, Kunos G. The endocannabinoid system as an emerging target of pharmacotherapy. Pharmacol Rev 2006;58:389–462.

[2] Devane WA, Hanus L, Breuer A, Pertwee RG, Stevenson LA, Griffin G, et al. Isolation and structure of a brain constituent that binds to the cannabinoid receptor. Science 1992;258:1946–9.

[3] Mechoulam R, Ben-Shabat S, Hanus L, Ligumsky M, Kaminski NE, Schatz AR, et al. Identification of an endogenous 2-monoglyceride, present in canine gut, that binds to cannabinoid receptors. Biochem Pharmacol 1995;50:83–90.

[4] Sugiura T, Kondo S, Sukagawa A, Nakane S, Shinoda A, Itoh K, et al. 2-Arachidonoylglycerol: a possible endogenous cannabinoid receptor ligand in brain. Biochem Biophys Res Commun 1995;215:89–97.

[5] Sugiura T, Kishimoto S, Oka S, Gokoh M. Biochemistry, pharmacology and physiology of 2-arachidonoylglycerol, an endogenous cannabinoid receptor ligand. Prog Lipid Res 2006;45:405–46.

[6] Pavón FJ, Serrano A, Romero-Cuevas M, Alonso M, Rodríguez de Fonseca F. Oleoyletha-nolamide: a new player in peripheral control of energy metabolism. Therapeutic implications. Drug Discov Today: Dis Mech 2010;7:e175–83.

[7] Hesselink JM. Evolution in pharmacologic thinking around the natural analgesic palmitoyl-ethanolamide: from nonspecific resistance to PPAR-α agonist and effective nutraceutical. J Pain Res 2013;6:625–34.

[8] Schmid HHO, Schmid PC, Natarajan V. N-Acylated glycerophospholipids and their deriva-tives. Prog Lipid Res 1990;29:1–43.

[9] Ueda N, Tsuboi K, Uyama T. Metabolism of endocannabinoids and related N-acylethanolamines: canonical and alternative pathways. FEBS J 2013;280:1874–94.

[10] Coulon D, Faure L, Salmon M, Wattelet V, Bessoule JJ. Occurrence, biosynthesis and functions of N-acylphosphatidylethanolamines (NAPE): not just precursors of N-acylethanolamines (NAE). Biochimie 2012;94:75–85.

[11] Hansen HS, Moesgaard B, Hansen HH, Petersen G. N-Acylethanolamines and precursor phospholipids – relation to cell injury. Chem Phys Lipids 2000;108:135–50.

[12] Shinohara N, Uyama T, Jin X-H, Tsuboi K, Tonai T, Houchi H, et al. Enzymological analysis of the tumor suppressor A-C1 reveals a novel group of phospholipid-metabolizing enzymes. J Lipid Res 2011;52:1927–35.

[13] Uyama T, Ikematsu N, Inoue M, Shinohara N, Jin X-H, Tsuboi K, et al. Generation of N-acylphosphatidylethanolamine by members of the phospholipase A/acyltransferase (PLA/AT) family. J Biol Chem 2012;287:31905–19.

[14] Uyama T, Inoue M, Okamoto Y, Shinohara N, Tai T, Tsuboi K, et al. Involvement of phos-pholipase A/acyltransferase-1 in N-acylphosphatidylethanolamine generation. Biochim Bio-phys Acta 2013;1831:1690–701.

[15] Uyama T, Ichi I, Kono N, Inoue A, Tsuboi K, Jin X-H, et al. Regulation of peroxisomal lipid metabolism by catalytic activity of tumor suppressor H-rev107. J Biol Chem 2012;287:2706–18.

[16] Okamoto Y, Morishita J, Tsuboi K, Tonai T, Ueda N. Molecular characterization of a phos-pholipase D generating anandamide and its congeners. J Biol Chem 2004;279:5298–305.

[17] Wang J, Okamoto Y, Morishita J, Tsuboi K, Miyatake A, Ueda N. Functional analysis of the purified anandamide-generating phospholipase D as a member of the metallo-β-lactamase family. J Biol Chem 2006;281:12325–35.

[18] Leung D, Saghatelian A, Simon GM, Cravatt BF. Inactivation of N-acyl phosphatidyletha-nolamine phospholipase D reveals multiple mechanisms for the biosynthesis of endocannabi-noids. Biochemistry 2006;45:4720–6.

[19] Tsuboi K, Okamoto Y, Ikematsu N, Inoue M, Shimizu Y, Uyama T, et al. Enzymatic formation of N-acylethanolamines from N-acylethanolamine plasmalogen through N-acylphosphatidylethanolamine-hydrolyzing phospholipase D-dependent and -independent pathways. Biochim Biophys Acta 2011;1811:565–77.

[20] Guo L, Gragg SD, Chen Z, Zhang Y, Amarnath V, Davies SS. Isolevuglandin-modified phos-phatidylethanolamine is metabolized by NAPE-hydrolyzing phospholipase D. J Lipid Res 2013;54:3151–7.

[21] Ueda N, Liu Q, Yamanaka K. Marked activation of the N-acylphosphatidylethanolamine-hydrolyzing phosphodiesterase by divalent cations. Biochim Biophys Acta 2001;1532:121–7.

[22] Zhu C, Solorzano C, Sahar S, Realini N, Fung E, Sassone-Corsi P, et al. Proinflammatory stimuli control N-acylphosphatidylethanolamine-specific phospholipase D expression in macrophages. Mol Pharmacol 2011;79:786–92.

[23] Wangensteen T, Akselsen H, Holmen J, Undlien D, Retterstøl L. A common haplotype in NAPEPLD is associated with severe obesity in a Norwegian population-based cohort (the HUNT study). Obesity 2011;19:612–7.

[24] Natarajan V, Schmid PC, Reddy PV, Schmid HHO. Catabolism of N-acylethanolamine phospholipids by dog brain preparations. J Neurochem 1984;42:1613–9.

[25] Sun Y-X, Tsuboi K, Okamoto Y, Tonai T, Murakami M, Kudo I, et al. Biosynthesis of anandamide and N-palmitoylethanolamine by sequential actions of phospholipase A_2 and lysophospholipase D. Biochem J 2004;380:749–56.

[26] Simon GM, Cravatt BF. Endocannabinoid biosynthesis proceeding through glycerophospho-N-acyl ethanolamine and a role for α/β-hydrolase 4 in this pathway. J Biol Chem 2006;281: 26465–72.

[27] Simon GM, Cravatt BF. Anandamide biosynthesis catalyzed by the phosphodiesterase GDE1 and detection of glycerophospho-N-acyl ethanolamine precursors in mouse brain. J Biol Chem 2008;283:9341–9.

[28] Simpson CD, Hurren R, Kasimer D, MacLean N, Eberhard Y, Ketela T, et al. A genome wide shRNA screen identifies α/β hydrolase domain containing 4 (ABHD4) as a novel regulator of anoikis resistance. Apoptosis 2012;17:666–78.

[29] Zheng B, Chen D, Farquhar MG. MIR16, a putative membrane glycerophosphodiester phosphodiesterase, interacts with RGS16. Proc Natl Acad Sci USA 2000;97:3999–4004.

[30] Zheng B, Berrie CP, Corda D, Farquhar MG. GDE1/MIR16 is a glycerophosphoinositol phosphodiesterase regulated by stimulation of G protein-coupled receptors. Proc Natl Acad Sci USA 2003;100:1745–50.

[31] Simon GM, Cravatt BF. Characterization of mice lacking candidate N-acyl ethanolamine biosynthetic enzymes provides evidence for multiple pathways that contribute to endocannabinoid production in vivo. Mol BioSyst 2010;6:1411–8.

[32] Kopp F, Komatsu T, Nomura DK, Trauger SA, Thomas JR, Siuzdak G, et al. The glycerophospho metabolome and its influence on amino acid homeostasis revealed by brain metabolomics of GDE1(−/−) mice. Chem Biol 2010;17:831–40.

[33] Astarita G, Ahmed F, Piomelli D. Identification of biosynthetic precursors for the endocannabinoid anandamide in the rat brain. J Lipid Res 2008;49:48–57.

[34] Liu J, Wang L, Harvey-White J, Osei-Hyiaman D, Razdan R, Gong Q, et al. A biosynthetic pathway for anandamide. Proc Natl Acad Sci USA 2006;103:13345–50.

[35] Liu J, Wang L, Harvey-White J, Huang BX, Kim HY, Luquet S, et al. Multiple pathways involved in the biosynthesis of anandamide. Neuropharmacology 2008;54:1–7.

[36] Cravatt BF, Demarest K, Patricelli M, Bracey MH, Giang DK, Martin BR, et al. Supersensitivity to anandamide and enhanced endogenous cannabinoid signaling in mice lacking fatty acid amide hydrolase. Proc Natl Acad Sci USA 2001;98:9371–6.

[37] McKinney MK, Cravatt BF. Structure and function of fatty acid amide hydrolase. Annu Rev Biochem 2005;74:411–32.

[38] Blankman JL, Cravatt BF. Chemical probes of endocannabinoid metabolism. Pharmacol Rev 2013;65:849–71.

[39] Bornheim LM, Kim KY, Chen B, Correia MA. The effect of cannabidiol on mouse hepatic microsomal cytochrome P450-dependent anandamide metabolism. Biochem Biophys Res Commun 1993;197:740–6.

[40] Ueda N, Yamamoto K, Yamamoto S, Tokunaga T, Shirakawa E, Shinkai H, et al. Lipoxygenase-catalyzed oxygenation of arachidonylethanolamide, a cannabinoid receptor agonist. Biochim Biophys Acta 1995;1254:127–34.

[41] Hampson AJ, Hill WAG, Zan-Phillips M, Makriyannis A, Leung E, Eglen RM, et al. Anandamide hydroxylation by brain lipoxygenase: metabolite structures and potencies at the cannabinoid receptor. Biochim Biophys Acta 1995;1259:173–9.

[42] Yu M, Ives D, Ramesha CS. Synthesis of prostaglandin E_2 ethanolamide from anandamide by cyclooxygenase-2. J Biol Chem 1997;272:21181–6.

[43] Kozak KR, Crews BC, Morrow JD, Wang LH, Ma YH, Weinander R, et al. Metabolism of the endocannabinoids, 2-arachidonylglycerol and anandamide, into prostaglandin, thromboxane, and prostacyclin glycerol esters and ethanolamides. J Biol Chem 2002;277:44877–85.

[44] Cravatt BF, Giang DK, Mayfield SP, Boger DL, Lerner RA, Gilula NB. Molecular characterization of an enzyme that degrades neuromodulatory fatty-acid amides. Nature 1996;384:83–7.

[45] Katayama K, Ueda N, Kurahashi Y, Suzuki H, Yamamoto S, Kato I. Distribution of anandamide amidohydrolase in rat tissues with special reference to small intestine. Biochim Biophys Acta 1997;1347:212–8.

[46] Wei BQ, Mikkelsen TS, McKinney MK, Lander ES, Cravatt BF. A second fatty acid amide hydrolase with variable distribution among placental mammals. J Biol Chem 2006;281:36569–78.

[47] McKinney MK, Cravatt BF. Evidence for distinct roles in catalysis for residues of the serine–serine–lysine catalytic triad of fatty acid amide hydrolase. J Biol Chem 2003;278:37393–9.

[48] Labar G, Michaux C. Fatty acid amide hydrolase: from characterization to therapeutics. Chem Biodivers 2007;4:1882–902.

[49] Bracey MH, Hanson MA, Masuda KR, Stevens RC, Cravatt BF. Structural adaptations in a membrane enzyme that terminates endocannabinoid signaling. Science 2002;298:1793–6.

[50] Ueda N, Kurahashi Y, Yamamoto S, Tokunaga T. Partial purification and characterization of the porcine brain enzyme hydrolyzing and synthesizing anandamide. J Biol Chem 1995;270:23823–7.

[51] Saghatelian A, Trauger SA, Want EJ, Hawkins EG, Siuzdak G, Cravatt BF. Assignment of endogenous substrates to enzymes by global metabolite profiling. Biochemistry 2004;43:14332–9.

[52] Goparaju SK, Ueda N, Yamaguchi H, Yamamoto S. Anandamide amidohydrolase reacting with 2-arachidonoylglycerol, another cannabinoid receptor ligand. FEBS Lett 1998;422:69–73.

[53] Blankman JL, Simon GM, Cravatt BF. A comprehensive profile of brain enzymes that hydrolyze the endocannabinoid 2-arachidonoylglycerol. Chem Biol 2007;14:1347–56.

[54] Sipe JC, Chiang K, Gerber AL, Beutler E, Cravatt BF. A missense mutation in human fatty acid amide hydrolase associated with problem drug use. Proc Natl Acad Sci USA 2002;99:8394–9.

[55] Fu J, Bottegoni G, Sasso O, Bertorelli R, Rocchia W, Masetti M, et al. A catalytically silent FAAH-1 variant drives anandamide transport in neurons. Nat Neurosci 2011;15:64–9.

[56] Leung K, Elmes MW, Glaser ST, Deutsch DG, Kaczocha M. Role of FAAH-like anandamide transporter in anandamide inactivation. PLoS One 2013;8:e79355.

[57] Liu J, Cinar R, Xiong K, Godlewski G, Jourdan T, Lin Y, et al. Monounsaturated fatty acids generated via stearoyl CoA desaturase-1 are endogenous inhibitors of fatty acid amide hydrolase. Proc Natl Acad Sci USA 2013;110:18832–7.

[58] Gulyas AI, Cravatt BF, Bracey MH, Dinh TP, Piomelli D, Boscia F, et al. Segregation of two endocannabinoid-hydrolyzing enzymes into pre- and postsynaptic compartments in the rat hippocampus, cerebellum and amygdala. Eur J Neurosci 2004;20:441–58.

[59] Kaczocha M, Glaser ST, Chae J, Brown DA, Deutsch DG. Lipid droplets are novel sites of N-acylethanolamine inactivation by fatty acid amide hydrolase-2. J Biol Chem 2010;285:2796–806.

[60] Kathuria S, Gaetani S, Fegley D, Valiño F, Duranti A, Tontini A, et al. Modulation of anxiety through blockade of anandamide hydrolysis. Nat Med 2003;9:76–81.

[61] Lichtman AH, Leung D, Shelton CC, Saghatelian A, Hardouin C, Boger DL, et al. Reversible inhibitors of fatty acid amide hydrolase that promote analgesia: evidence for an unprecedented combination of potency and selectivity. J Pharmacol Exp Ther 2004;311:441–8.

[62] Ahn K, Johnson DS, Mileni M, Beidler D, Long JZ, McKinney MK, et al. Discovery and characterization of a highly selective FAAH inhibitor that reduces inflammatory pain. Chem Biol 2009;16:411–20.

[63] Ahn K, Smith SE, Liimatta MB, Beidler D, Sadagopan N, Dudley DT, et al. Mechanistic and pharmacological characterization of PF-04457845: a highly potent and selective fatty acid amide hydrolase inhibitor that reduces inflammatory and noninflammatory pain. J Pharmacol Exp Ther 2011;338:114–24.

[64] Nomura DK, Blankman JL, Simon GM, Fujioka K, Issa RS, Ward AM, et al. Activation of the endocannabinoid system by organophosphorus nerve agents. Nat Chem Biol 2008;4:373–8.

[65] Long JZ, Nomura DK, Vann RE, Walentiny DM, Booker L, Jin X, et al. Dual blockade of FAAH and MAGL identifies behavioral processes regulated by endocannabinoid crosstalk in vivo. Proc Natl Acad Sci USA 2009;106:20270–5.

[66] Ueda N, Yamanaka K, Terasawa Y, Yamamoto S. An acid amidase hydrolyzing anandamide as an endogenous ligand for cannabinoid receptors. FEBS Lett 1999;454:267–70.

[67] Ueda N, Yamanaka K, Yamamoto S. Purification and characterization of an acid amidase selective for N-palmitoylethanolamine, a putative endogenous anti-inflammatory substance. J Biol Chem 2001;276:35552–7.

[68] Tsuboi K, Sun Y-X, Okamoto Y, Araki N, Tonai T, Ueda N. Molecular characterization of N-acylethanolamine-hydrolyzing acid amidase, a novel member of the choloylglycine hydrolase family with structural and functional similarity to acid ceramidase. J Biol Chem 2005;280:11082–92.

[69] Tsuboi K, Takezaki N, Ueda N. The N-acylethanolamine-hydrolyzing acid amidase (NAAA). Chem Biodivers 2007;4:1914–25.

[70] Ueda N, Tsuboi K, Uyama T. N-Acylethanolamine metabolism with special reference to N-acylethanolamine-hydrolyzing acid amidase (NAAA). Prog Lipid Res 2010;49:299–315.

[71] Zhao L-Y, Tsuboi K, Okamoto Y, Nagahata S, Ueda N. Proteolytic activation and glycosylation of N-acylethanolamine-hydrolyzing acid amidase, a lysosomal enzyme involved in the endocannabinoid metabolism. Biochim Biophys Acta 2007;1771:1397–405.

[72] West JM, Zvonok N, Whitten KM, Wood JT, Makriyannis A. Mass spectrometric characterization of human N-acylethanolamine-hydrolyzing acid amidase. J Proteome Res 2012;11:972–81.

[73] Hong S-B, Li C-M, Rhee H-J, Park J-H, He X, Levy B, et al. Molecular cloning and characterization of a human cDNA and gene encoding a novel acid ceramidase-like protein. Genomics 1999;62:232–41.

[74] Armirotti A, Romeo E, Ponzano S, Mengatto L, Dionisi M, Karacsonyi C, et al. β-Lactones inhibit N-acylethanolamine acid amidase by S-acylation of the catalytic N-terminal cysteine. ACS Med Chem Lett 2012;3:422–6.

[75] West JM, Zvonok N, Whitten KM, Vadivel SK, Bowman AL, Makriyannis A. Biochemical and mass spectrometric characterization of human N-acylethanolamine-hydrolyzing acid amidase inhibition. PLoS One 2012;7:e43877.

[76] Tai T, Tsuboi K, Uyama T, Masuda K, Cravatt BF, Houchi H, et al. Endogenous molecules stimulating N-acylethanolamine-hydrolyzing acid amidase (NAAA). ACS Chem Neurosci 2012;3:379–85.

[77] Sun Y-X, Tsuboi K, Zhao L-Y, Okamoto Y, Lambert DM, Ueda N. Involvement of N-acylethanolamine-hydrolyzing acid amidase in the degradation of anandamide and other N-acylethanolamines in macrophages. Biochim Biophys Acta 2005;1736:211–20.

[78] Tsuboi K, Zhao L-Y, Okamoto Y, Araki N, Ueno M, Sakamoto H, et al. Predominant expression of lysosomal N-acylethanolamine-hydrolyzing acid amidase in macrophages revealed by immunochemical studies. Biochim Biophys Acta 2007;1771:623–32.

[79] Wang J, Zhao L-Y, Uyama T, Tsuboi K, Wu X-X, Kakehi Y, et al. Expression and secretion of N-acylethanolamine-hydrolysing acid amidase in human prostate cancer cells. J Biochem 2008;144:685–90.

[80] Petrosino S, Iuvone T, Di Marzo V. N-Palmitoyl-ethanolamine: biochemistry and new therapeutic opportunities. Biochimie 2010;92:724–7.

[81] Solorzano C, Zhu C, Battista N, Astarita G, Lodola A, Rivara S, et al. Selective N-acylethanolamine-hydrolyzing acid amidase inhibition reveals a key role for endogenous palmitoylethanolamide in inflammation. Proc Natl Acad Sci USA 2009;106:20966–71.

[82] Solorzano C, Antonietti F, Duranti A, Tontini A, Rivara S, Lodola A, et al. Synthesis and structure-activity relationships of N-(2-oxo-3-oxetanyl)amides as N-acylethanolamine-hydrolyzing acid amidase inhibitors. J Med Chem 2010;53:5770–81.

[83] Duranti A, Tontini A, Antonietti F, Vacondio F, Fioni A, Silva C, et al. N-(2-Oxo-3-oxetanyl) carbamic acid esters as N-acylethanolamine acid amidase inhibitors: synthesis and structure–activity and structure–property relationships. J Med Chem 2012;55:4824–36.

[84] Ponzano S, Bertozzi F, Mengatto L, Dionisi M, Armirotti A, Romeo E, et al. Synthesis and structure–activity relationship (SAR) of 2-methyl-4-oxo-3-oxetanylcarbamic acid esters, a class of potent N-acylethanolamine acid amidase (NAAA) inhibitors. J Med Chem 2013;56:6917–34.

[85] Sasso O, Moreno-Sanz G, Martucci C, Realini N, Dionisi M, Mengatto L, et al. Antinociceptive effects of the N-acylethanolamine acid amidase inhibitor ARN077 in rodent pain models. Pain 2013;154:350–60.

[86] Sugiura T, Kishimoto S, Oka S, Gokoh M. Biochemistry, pharmacology and physiology of 2-arachidonoylglycerol, an endogenous cannabinoid receptor ligand. Prog Lipid Res 2006;45:405–46.

[87] Kano M, Ohno-Shosaku T, Hashimotodani Y, Uchigashima M, Watanabe M. Endocannabinoid-mediated control of synaptic transmission. Physiol Rev 2009;89:309–80.

[88] Murataeva N, Straiker A, Mackie K. Parsing the players: 2-arachidonoylglycerol synthesis and degradation in the CNS. Br J Pharmacol 2014;171:1379–91.

[89] Fukami K, Inanobe S, Kanemaru K, Nakamura Y. Phospholipase C is a key enzyme regulating intracellular calcium and modulating the phosphoinositide balance. Prog Lipid Res 2010;49:429–37.

[90] Hashimotodani Y, Ohno-Shosaku T, Tsubokawa H, Ogata H, Emoto K, Maejima T, et al. Phospholipase Cβ serves as a coincidence detector through its Ca^{2+} dependency for triggering retrograde endocannabinoid signal. Neuron 2005;45:257–68.

[91] Maejima T, Oka S, Hashimotodani Y, Ohno-Shosaku T, Aiba A, Wu D, et al. Synaptically driven endocannabinoid release requires Ca^{2+}-assisted metabotropic glutamate receptor subtype 1 to phospholipase Cβ4 signaling cascade in the cerebellum. J Neurosci 2005;25: 6826–35.

[92] Bisogno T, Melck D, De Petrocellis L, Di Marzo V. Phosphatidic acid as the biosynthetic precursor of the endocannabinoid 2-arachidonoylglycerol in intact mouse neuroblastoma cells stimulated with ionomycin. J Neurochem 1999;72:2113–9.

[93] Oka S, Yanagimoto S, Ikeda S, Gokoh M, Kishimoto S, Waku K, et al. Evidence for the involvement of the cannabinoid CB2 receptor and its endogenous ligand 2-arachidonoylglycerol in 12-O-tetradecanoylphorbol-13-acetate-induced acute inflammation in mouse ear. J Biol Chem 2005;280:18488–97.

[94] Ueda H, Kobayashi T, Kishimoto M, Tsutsumi T, Okuyama H. A possible pathway of phosphoinositide metabolism through EDTA-insensitive phospholipase A$_1$ followed by lysophosphoinositide-specific phospholipase C in rat brain. J Neurochem 1993;61:1874–81.

[95] Nakane S, Oka S, Arai S, Waku K, Ishima Y, Tokumura A, et al. 2-Arachidonoyl-*sn*-glycero-3-phosphate, an arachidonic acid-containing lysophosphatidic acid: occurrence and rapid enzymatic conversion to 2-arachidonoyl-*sn*-glycerol, a cannabinoid receptor ligand, in rat brain. Arch Biochem Biophys 2002;402:51–8.

[96] Bisogno T, Howell F, Williams G, Minassi A, Cascio MG, Ligresti A, et al. Cloning of the first *sn*1-DAG lipases points to the spatial and temporal regulation of endocannabinoid signaling in the brain. J Cell Biol 2003;163:463–8.

[97] Oudin MJ, Hobbs C, Doherty P. DAGL-dependent endocannabinoid signalling: roles in axonal pathfinding, synaptic plasticity and adult neurogenesis. Eur J Neurosci 2011;34:1634–46.

[98] Jung KM, Astarita G, Zhu C, Wallace M, Mackie K, Piomelli D. A key role for diacylglycerol lipase-alpha in metabotropic glutamate receptor-dependent endocannabinoid mobilization. Mol Pharmacol 2007;72:612–21.

[99] Tanimura A, Yamazaki M, Hashimotodani Y, Uchigashima M, Kawata S, Abe M, et al. The endocannabinoid 2-arachidonoylglycerol produced by diacylglycerol lipase α mediates retrograde suppression of synaptic transmission. Neuron 2010;65:320–7.

[100] Gao Y, Vasilyev DV, Goncalves MB, Howell FV, Hobbs C, Reisenberg M, et al. Loss of retrograde endocannabinoid signaling and reduced adult neurogenesis in diacylglycerol lipase knock-out mice. J Neurosci 2010;30:2017–24.

[101] Jain T, Wager-Miller J, Mackie K, Straiker A. Diacylglycerol lipase α (DAGLα) and DAGLβ cooperatively regulate the production of 2-arachidonoyl glycerol in autaptic hippocampal neurons. Mol Pharmacol 2013;84:296–302.

[102] Jeong WI, Osei-Hyiaman D, Park O, Liu J, Bátkai S, Mukhopadhyay P, et al. Paracrine activation of hepatic CB1 receptors by stellate cell-derived endocannabinoids mediates alcoholic fatty liver. Cell Metab 2008;7:227–35.

[103] Shonesy BC, Wang X, Rose KL, Ramikie TS, Cavener VS, Rentz T, et al. CaMKII regulates diacylglycerol lipase-α and striatal endocannabinoid signaling. Nat Neurosci 2013;16:456–63.

[104] Hoover HS, Blankman JL, Niessen S, Cravatt BF. Selectivity of inhibitors of endocannabinoid biosynthesis evaluated by activity-based protein profiling. Bioorg Med Chem Lett 2008;18:5838–41.

[105] Bisogno T, Cascio MG, Saha B, Mahadevan A, Urbani P, Minassi A, et al. Development of the first potent and specific inhibitors of endocannabinoid biosynthesis. Biochim Biophys Acta 2006;1761:205–12.

[106] Bisogno T, Burston JJ, Rai R, Allarà M, Saha B, Mahadevan A, et al. Synthesis and pharmacological activity of a potent inhibitor of the biosynthesis of the endocannabinoid 2-arachidonoylglycerol. ChemMedChem 2009;4:946–50.

[107] Kanoh H, Iwata T, Ono T, Suzuki T. Immunological characterization of *sn*-1,2-diacylglycerol and *sn*-2-monoacylglycerol kinase from pig brain. J Biol Chem 1986;261:5597–602.

[108] Coleman RA, Haynes EB. Monoacylglycerol acyltransferase. Evidence that the activities from rat intestine and suckling liver are tissue-specific isoenzymes. J Biol Chem 1986;261:224–8.

[109] Kozak KR, Marnett LJ. Oxidative metabolism of endocannabinoids. Prostaglandins Leukot Essent Fatty Acids 2002;66:211–20.

[110] Rouzer CA, Marnett LJ. Non-redundant functions of cyclooxygenases: oxygenation of endocannabinoids. J Biol Chem 2008;283:8065–9.

[111] Labar G, Wouters J, Lambert DM. A review on the monoacylglycerol lipase: at the interface between fat and endocannabinoid signalling. Curr Med Chem 2010;17:2588–607.

[112] Karlsson M, Contreras JA, Hellman U, Tornqvist H, Holm C. cDNA cloning, tissue distribution, and identification of the catalytic triad of monoglyceride lipase. Evolutionary relationship to esterases, lysophospholipases, and haloperoxidases. J Biol Chem 1997;272:27218–23.

[113] Karlsson M, Reue K, Xia YR, Lusis AJ, Langin D, Tornqvist H, et al. Exon–intron organization and chromosomal localization of the mouse monoglyceride lipase gene. Gene 2001;272:11–8.

[114] Dinh TP, Carpenter D, Leslie FM, Freund TF, Katona I, Sensi SL, et al. Brain monoglyceride lipase participating in endocannabinoid inactivation. Proc Natl Acad Sci USA 2002;99:10819–24. and Erratum 2002;99:13961.

[115] Labar G, Bauvois C, Borel F, Ferrer JL, Wouters J, Lambert DM. Crystal structure of the human monoacylglycerol lipase, a key actor in endocannabinoid signaling. Chembiochem 2010;11:218–27.

[116] Dinh TP, Kathuria S, Piomelli D. RNA interference suggests a primary role for monoacylglycerol lipase in the degradation of the endocannabinoid 2-arachidonoylglycerol. Mol Pharmacol 2004;66:1260–4.

[117] Schlosburg JE, Blankman JL, Long JZ, Nomura DK, Pan B, Kinsey SG, et al. Chronic monoacylglycerol lipase blockade causes functional antagonism of the endocannabinoid system. Nat Neurosci 2010;13:1113–9.

[118] Chanda PK, Gao Y, Mark L, Btesh J, Strassle BW, Lu P, et al. Monoacylglycerol lipase activity is a critical modulator of the tone and integrity of the endocannabinoid system. Mol Pharmacol 2010;78:996–1003.

[119] Prescott SM, Majerus PW. Characterization of 1,2-diacylglycerol hydrolysis in human platelets. Demonstration of an arachidonoyl-monoacylglycerol intermediate. J Biol Chem 1983;258:764–9.

[120] Taschler U, Radner FP, Heier C, Schreiber R, Schweiger M, Schoiswohl G, et al. Monoglyceride lipase deficiency in mice impairs lipolysis and attenuates diet-induced insulin resistance. J Biol Chem 2011;286:17467–77.

[121] Long JZ, Li W, Booker L, Burston JJ, Kinsey SG, Schlosburg JE, et al. Selective blockade of 2-arachidonoylglycerol hydrolysis produces cannabinoid behavioral effects. Nat Chem Biol 2009;5:37–44.

[122] Nomura DK, Morrison BE, Blankman JL, Long JZ, Kinsey SG, Marcondes MC, et al. Endocannabinoid hydrolysis generates brain prostaglandins that promote neuroinflammation. Science 2011;334:809–13.

[123] Nomura DK, Long JZ, Niessen S, Hoover HS, Ng SW, Cravatt BF. Monoacylglycerol lipase regulates a fatty acid network that promotes cancer pathogenesis. Cell 2010;140:49–61.

[124] Petrosino S, Di Marzo V. FAAH and MAGL inhibitors: therapeutic opportunities from regulating endocannabinoid levels. Curr Opin Investig Drugs 2010;11:51–62.

[125] Minkkilä A, Saario S, Nevalainen T. Discovery and development of endocannabinoid-hydrolyzing enzyme inhibitors. Curr Top Med Chem 2010;10:828–58.

[126] Fowler CJ. Monoacylglycerol lipase – a target for drug development? Br J Pharmacol 2012;166:1568–85.

[127] Hohmann AG, Suplita RL, Bolton NM, Neely MH, Fegley D, Mangieri R, et al. An endocannabinoid mechanism for stress-induced analgesia. Nature 2005;435:1108–12.

[128] Muccioli GG, Labar G, Lambert DM. CAY10499, a novel monoglyceride lipase inhibitor evidenced by an expeditious MGL assay. Chembiochem 2008;9:2704–10.

[129] Bisogno T, Ortar G, Petrosino S, Morera E, Palazzo E, Nalli M, et al. Development of a potent inhibitor of 2-arachidonoylglycerol hydrolysis with antinociceptive activity in vivo. Biochim Biophys Acta 2009;1791:53–60.

[130] Chang JW, Niphakis MJ, Lum KM, Cognetta AB 3rd, Wang C, Matthews ML, et al. Highly selective inhibitors of monoacylglycerol lipase bearing a reactive group that is bioisosteric with endocannabinoid substrates. Chem Biol 2012;19:579–88.

[131] Ghosh S, Wise LE, Chen Y, Gujjar R, Mahadevan A, Cravatt BF, et al. The monoacylglycerol lipase inhibitor JZL184 suppresses inflammatory pain in the mouse carrageenan model. Life Sci 2013;92:498–505.

[132] Savinainen JR, Saario SM, Laitinen JT. The serine hydrolases MAGL, ABHD6 and ABHD12 as guardians of 2-arachidonoylglycerol signalling through cannabinoid receptors. Acta Physiol 2012;204:267–76. and *Erratum* 2012;204:460.

[133] Navia-Paldanius D, Savinainen JR, Laitinen JT. Biochemical and pharmacological characterization of human α/β-hydrolase domain containing 6 (ABHD6) and 12 (ABHD12). J Lipid Res 2012;53:2413–24.

[134] Marrs WR, Blankman JL, Horne EA, Thomazeau A, Lin YH, Coy J, et al. The serine hydrolase ABHD6 controls the accumulation and efficacy of 2-AG at cannabinoid receptors. Nat Neurosci 2010;13:951–7.

[135] Marrs WR, Horne EA, Ortega-Gutierrez S, Cisneros JA, Xu C, Lin YH, et al. Dual inhibition of α/β-hydrolase domain 6 and fatty acid amide hydrolase increases endocannabinoid levels in neurons. J Biol Chem 2011;286:28723–8.

[136] Thomas G, Betters JL, Lord CC, Brown AL, Marshall S, Ferguson D, et al. The serine hydrolase ABHD6 is a critical regulator of the metabolic syndrome. Cell Rep 2013;5:508–20.

[137] Fiskerstrand T, H'mida-Ben Brahim D, Johansson S, M'zahem A, Haukanes BI, Drouot N, et al. Mutations in ABHD12 cause the neurodegenerative disease PHARC: an inborn error of endocannabinoid metabolism. Am J Hum Genet 2010;87:410–7.

[138] Blankman JL, Long JZ, Trauger SA, Siuzdak G, Cravatt BF. ABHD12 controls brain lysophosphatidylserine pathways that are deregulated in a murine model of the neurodegenerative disease PHARC. Proc Natl Acad Sci USA 2013;110:1500–5.

CHAPTER 9

Endocannabinoidomics: "Omics" Approaches Applied to Endocannabinoids and Endocannabinoid-Like Mediators

Fabiana Piscitelli

9.1 INTRODUCTION

The discovery of the cannabinoid receptors and their endogenous ligands opened the way to the study of the "endocannabinoid system" (ECS) and its involvement in the regulation of several physiological and pathological conditions. Since several studies have shown that the levels of the major endocannabinoids (ECs) undergo pronounced changes in

The Endocannabinoidome: The World of Endocannabinoids and Related Mediators. DOI: 10.1016/B978-0-12-420126-2.00009-2
Copyright © 2015 Elsevier Inc. All rights reserved

biological fluids, tissues, and cells following various physiological, pathological, and pharmacological stimuli [1], the development of accurate analytical methods that identify and quantify EC levels in biological matrices is an essential tool for the understanding of the role of these as well as of EC-like mediators.

ECs and their lipid congeners are all derivatives of fatty acids, and since increasing interest has been focused on the biological significance of these lipid mediators, several "lipidomics" approaches have been applied to identify new EC-related molecules and investigate new molecular pathways involved in their metabolism. In this scenario, the term "endocannabinoidomics" was coined [2] to define the methodologies necessary to investigate the metabolomic, proteomic, and genomic components of the "endocannabinoidome." Targeted lipidomic methods to detect novel bioactive fatty acid derivatives as well as monitor known EC-like molecules and their biosynthetic precursors simultaneously, have been developed. Several of these EC congeners exert important physiological or pathological roles independently from cannabinoid receptors.

9.2 BIOCHEMISTRY AND PHARMACOLOGY OF ENDOCANNABINOIDS AND ENDOCANNABINOID-RELATED COMPOUNDS

Each EC and EC-like molecule could be classified on the basis of the lipid class to which they belong (Figure 9.1): (1) N-acylethanolamines (NAEs); (2) monoacylglycerols (MAGs); (3) N-acyldopamines; (4) fatty acid amides of amino acids (FAAAs or lipoaminoacids); (5) COX-2 derivatives; and (6) N-acylserotonins.

NAEs, the ethanolamides of long-chain fatty acids, are a class of naturally occurring lipid molecules with a variety of biological activities. Depending on the nature of the acyl chain, NAEs in animals can be involved in numerous physiological processes that have been deeply reviewed, such as in neuroprotection, neurotoxicity, cell proliferation, pain, inflammation, fertility, apoptosis, anxiety, cognition and memory, and food intake [3–6]. Even though N-arachidonoylethanolamide (anandamide or AEA) is the most studied NAE because of its activity at cannabinoid CB_1 and CB_2 receptors [4,7], it should be noted that this compound is a minor component in animal tissues, whereas other

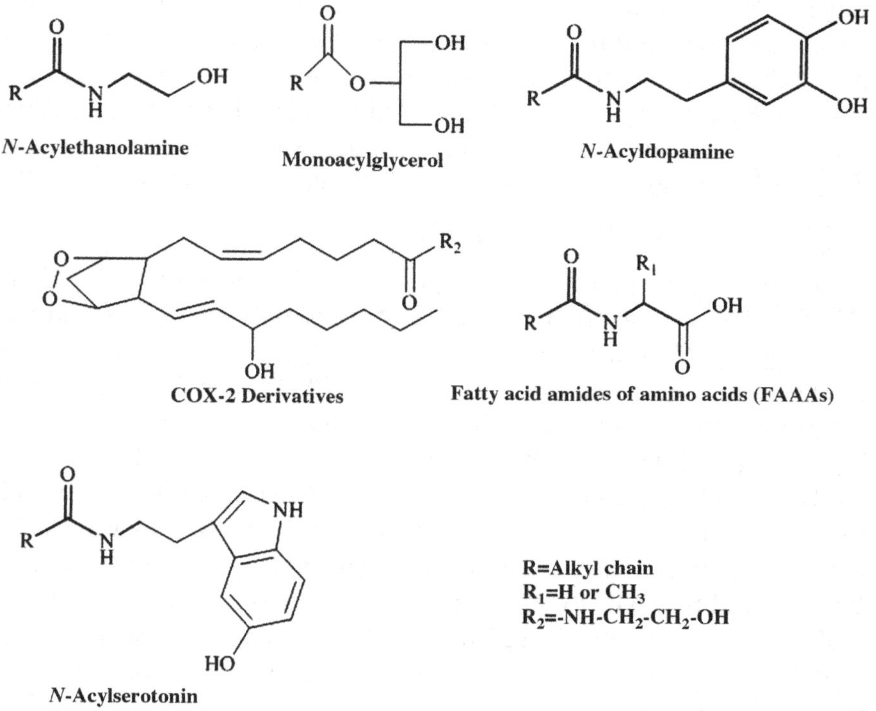

Fig. 9.1. Chemical structures of ECs and EC-like molecules.

NAEs, such as *N*-palmitoylethanolamide (PEA), stearoylethanolamide, *N*-oleoylethanolamide (OEA), and linoleoylethanolamide, are more abundant [8]. PEA and OEA, in particular, are strictly correlated to the ECS because the endogenous levels of these two lipid mediators have been often found to change together with EC levels in several diseases, such as metabolic disorders [9]. Although not interacting directly with cannabinoid receptors, they share the same biosynthetic pathways and are able to activate, directly or indirectly, several targets, such as the peroxisome proliferator–activated receptor-α (PPARα) [10] and the transient receptor potential vanilloid type-1 (TRPV1) channel [11–14]. Furthermore, evidence exists suggesting that PEA and OEA also activate GPR55 and GPR119 [15,16].

Three years after the discovery of AEA, a second endogenous ligand for cannabinoid receptors named 2-arachidonoylglycerol (2-AG) was identified [17,18]. Although both ECs possess an arachidonic acid chain in their structure, 2-AG belongs to the MAG lipid class and has metabolic pathways completely different from NAEs in animal tissues [19–21]. By analogy to the NAEs, MAGs other than 2-AG may also show cannabinoid receptor-independent activities. In fact, unsaturated 2-acylglycerols, namely 2-oleoyl- and 2-linoleoyl-glycerol (2-LG), also activate GPR119 [22].

Among long chain *N*-acylopamines, *N*-arachidonoyldopamine (NADA) is the most studied because of its activity at TRPV1 and CB$_1$ receptors [23,24] and its ability to inhibit T-type Ca^{2+} channels [25], as discussed in other chapters of this book. NADA was found in several rat brain regions with the highest concentrations in the striatum and hippocampus (max, ~6 pmol/g wet tissue weight). Another member of the *N*-acyldopamine family, the *N*-oleoyldopamine (OLDA), is able to activate TRPV1 and is much more selective than NADA versus CB$_1$ receptors [26]. Instead, other endogenous *N*-acyldopamines, termed PALDA and STEARDA (the dopamine amides of palmitic and stearic acid, respectively), even though inactive *per se* on TRPV1, act *in vitro* as "entourage" compounds, by enhancing the effect of AEA or NADA on this receptor [26].

Regarding FAAAs, the first such compound to be discovered was *N*-arachidonoylglycine (NAGly) in 1997 [27]. At low concentrations, NAGly is able to activate GPR18, an orphan receptor coupled to G proteins [28]. Rimmerman et al. showed that *N*-palmitoylglycine (PalGly) is produced endogenously and plays a role in sensory neuronal signaling [29]. Further work from the late Walker's laboratory identified other endogenous *N*-acylglycines: *N*-oleoylglycine (OlGly), *N*-stearoylglycine (StrGly), *N*-linoleoylglycine (LinGly), and *N*-docosahexaenoylglycine (DocGly) [30]. The identification of these novel compounds led to the hypothesis that they have potential roles as signaling molecules, as precursors of other bioactive lipids, or both. Further investigations are needed to understand the biological roles of these lipids and their mechanism of action.

Multiple investigations into EC metabolic pathways have been carried out and accumulating data suggest that COX-2 derivatives act as good substrates for several fatty acid oxygenases, as discussed in other chapters. These oxygenases include the cyclooxygenase-2 (COX-2) [31], known to be involved in prostanoid production from arachidonic acid. The oxygenation of ECs by COX-2 leads to the formation of prostaglandin-ethanolamides (prostamides, PMs) and prostaglandin-glyceryl esters (PG-GEs) with unique pharmacological properties [31]. $PMF_{2\alpha}$ received the greatest attention together with its analog bimatoprost, an antiglaucoma drug, even though their action is not mediated by homomeric prostanoid (FP) receptors [32].

Recently, Verhoeckx et al. reported the endogenous formation of several N-acylserotonins in the gastrointestinal tract of pigs and mice [33]. In particular, they developed an LC/MS/MS method for the analysis of six different N-acylserotonins: N-palmitoylserotonin (PA-5-HT), N-stearoylserotonin (SA-5-HT), N-oleoylserotonin (OA-5-HT), N-arachidonoylserotonin (AA-5-HT), N-eicosapentaenoylserotonin (EPA-5-HT), and N-docosahexaenoylserotonin (DHA-5-HT); and they reported that PA-5-HT and OA-5-HT were the most prominent ones detected in the porcine intestine [33]. Moreover, N-acylserotonins, when polyunsaturated, inhibit FAAH and antagonize TRPV1 channels [34,35]. However, further investigations are needed to understand the possible biological significance of these metabolites.

9.3 LIPIDOMICS IN "ENDOCANNABINOIDOMICS": MASS-SPECTROMETRIC APPROACHES FOR ENDOCANNABINOID AND ENDOCANNABINOID-LIKE MOLECULE QUANTIFICATION

9.3.1 Analytical Approaches Available for Small Molecules of the Endocannabinoidome

Numerous analytical approaches to detect ECs and EC-related molecules have been proposed to date, most of which based on mass spectrometry in combination with liquid chromatography (LC/MS or LC/MS/MS) or gas chromatography (GC/MS or GC/MS/MS). In principle, the analytical technologies used to determine the chemical structure of anandamide (AEA) were [1]H-nuclear magnetic resonance (NMR) and GC/MS [36], but since then considerable progress has been made toward the development of more sensitive techniques. Despite the great improvement in instrumentation, sample extraction and purification, quantification of ECs and EC-related compounds remains an analytical challenge due to their low levels in biological matrices and their instability. Recently, Zoerner et al. reviewed all the analytical techniques and methodologies used to date, to quantify ECs from biological samples [37]. This overview goes from tissue extraction and purification techniques (Folch method, Bligh and Dyer method, liquid–liquid extraction, solid-phase extraction, and microdialysis techniques) to the analytical methods available to detect ECs [high-performance liquid chromatography with UV or fluorescence detection (HPLC-UV, HPLC-FL), ultra-high-performance liquid chromatography (UPLC/MS), GC/MS, and GC/MS/MS]. It is noteworthy that most ECs and EC-like mediators are present in biological matrices in very low amounts and given the lack of chromophoric or fluorescent moieties in their molecules, EC analysis by HPLC with UV or fluorescence detection requires previous steps of derivatization, as with most of the GC/MS techniques. LC/MS and LC/MS/MS methodologies, however, provided the advantage of avoiding the derivatization and enabling the best analytical performance in terms of accuracy, sensitivity, and precision. In particular, the introduction of HPLC coupled to electrospray (LC/ESI/MS) or atmospheric pressure chemical ionization MS (LC/APCI/MS) provided the rapidity, best high-throughput capability, and subsequent miniaturization of the HPLC components for the detection of very low amounts of these compounds. Generally LC/MS methods for the detection of ECs are based on isotope dilution, which results in comparing the ratio mass/charge (m/z) of the deuterated standards to that of the "unknown" compound, using the selected ion monitoring (SIM) mode. A similar approach, but more sensitive than SIM and requiring the use of tandem mass spectrometry LC/MS/MS, is the selected reaction monitoring (SRM) mode, in which only a single ion fragment from a single precursor ion is detected; while in the multiple reaction monitoring (MRM), multiple product ions from one or more precursor ions are detected [2].

The development of ever more advanced and sophisticated LC/MS/MS techniques coincided with the concomitant identification of new EC-like molecules, underlining the emerging need for a more targeted approach. Usually, LC/ESI/MS or LC/ESI/MS/MS analysis of ECs, such as AEA and 2-AG, is performed in the positive ion mode, yielding three possible cations, i.e., for AEA: $[AEA + H]^+$, $[AEA + Na]^+$ [38,39] and $[AEA + K]^+$ [40] at m/z 348, 370, and 386, respectively. Collision-induced dissociation (CID) of the ion [AEA + H]$^+$ yields the characteristic product ion at m/z 62, corresponding to the protonated ethanolamide moiety ([OHCH$_2$CH$_2$NH$_3$]$^+$) [38]. Kingsley and Marnett proposed an interesting approach to increase sensitivity of AEA and 2-AG in LC/MS/MS analysis by adding silver acetate to the mobile phase [41]. In particular, the silver cation is able to bind the polyunsaturated arachidonate backbone of both molecules and allows the detection of AEA and 2-AG with a limit of detection (LOD) of 13 and 14 fmol, respectively [41]. In our laboratory, the use of a Shimadzu high-performance liquid chromatography apparatus (LC-10ADVP) coupled to a Shimadzu (LCMS-2020) quadrupole mass spectrometry via a Shimadzu atmospheric pressure chemical ionization interface for the simultaneous detection of AEA, 2-AG, PEA, and OEA from several biological matrices, provided a highly sensitive analysis (LOD = 10 fmol) [42]. Furthermore, we set up a method for the sensitive and accurate detection of NADA, as well as the simultaneous monitoring of the above-mentioned ECs, from neural precursor cells (NPSCs), and even with a LOD of 10 fmol, NADA was not detectable in any sample [43].

Increasing evidence, coming from several LC/MS/MS approaches [i.e., ion trap (IT) LC/MS/MS, quadrupole-time-of-flight (QToF) LC/MS/MS analysis], suggests that the superfamily of EC-like molecules is enlarging. In fact, these approaches have been applied in mammalian tissues to identify FAAAs, such as N-arachidonoyl γ-aminobutyric acid (NAGABA), N-arachidonoyl alanine (NAAla), and N-arachidonoyl-L-serine (NASer) [44,45]. However, even though several studies have been carried out to understand the effects of these metabolites *in vivo* or *in vitro* [vasodilatatory [46] and neuroprotective [47]], their mechanisms of action are yet to be elucidated. Moreover, recently, a triple quadrupole LC/MS/MS method has been applied to simultaneously detect and quantify the levels of NAGly, NAAla, NASer, NAGABA, AEA, and 2-AG in the mouse brain [48]. Very recently, another LC/MS/MS method was set up to measure AEA, 2-AG, PEA, OEA, and LEA (N-linoleoylethanolamide) in brain tissues, and the quantification limit reached 70 nmol/g and 0.3 mmol/g, for NAEs and 2-AG, respectively [49].

In the literature, there are several papers in which authors measure EC levels using different extraction and detection methods. In particular,

Zoerner et al. reported a sensitive UPLC/MS/MS method to quantify AEA, 2-AG, and 1-AG in human blood samples, using toluene extraction to avoid matrix-effects, 2-AG/1-AG isomerization and degradation [40]. Thieme et al., instead, reported the quantification of AEA and 2-AG in plasma samples of patients who received a large intravenous dose of tetrahydrocannabinol [50]. In this case, samples were extracted with ethyl acetate and cyclohexane, in the presence of deuterated internal standards, and analyzed by a triple quadrupole LC/MS/MS [50]. The same LC/MS/MS technology was used to simultaneously quantify AEA, 2-AG, PEA, OEA, LEA, N-stearoylethanolamine (SEA), N-docosahexaenoyl-ethanolamine (DHEA), N-dihomo-γ-linolenoyl-ethanolamine (DGLEA), and 2-LG in human plasma from abstinent cocaine-addicted subjects [51]. Regarding N-acylserotonins, very little is known about these mediators because only a few papers reported their endogenous formation. Nevertheless, the identification of endogenous AA-5-HT has opened the way to new fields of research. In fact, very recently, the identification of an oxidative product of AA-5-HT by cytochrome P450 in human brain tumor tissues was reported [52], using an UPLC coupled to an LTQ for isotope dilution experiments, Q-TOF to determine the accurate mass and, finally, a ^{1}H-NMR analysis to confirm the chemical structure.

9.3.2 High Resolution Ion-Trap/Time-of-Flight LC/MS for the Unequivocal Identification of New EC-Like Molecules and EC Biosynthetic Precursors

The application of very targeted methods to identify EC-like molecules is becoming necessary for the understanding of their biological role and the development of new diagnostic biomarkers. Our laboratory developed a novel analytical technique for the unequivocal identification and quantification of the major PMs and PG-GEs and tested the effects of one of these compounds, as well as of selective antagonists for its proposed receptor, on pain perception and dorsal horn nociceptive (NS) neuron hyperexcitability, in healthy and/or knee-inflamed mice [53]. In particular, EC COX-2 derivative quantification was performed by a high-resolution ion trap-time-of-flight mass spectrometry (LC/MS/IT/TOF) analysis (Shimadzu Corporation, Kyoto, Japan) equipped with an ESI interface, using MRM. This analysis provided high-resolution

[M + Na]$^+$ adducts, and PM and PG-GE quantification was performed by isotope dilution. CID of the ion [M + Na]$^+$ yielded the characteristic loss of water. The recovery of PME$_2$, PMF$_{2\alpha}$, PGE$_2$-GE, and PGF$_{2\alpha}$-GE from tissues was 42.5 ± 1.9%, 61.6 ± 15.9%, 49.1 ± 15.7%, and 52.3 ± 17.8%, respectively. The LC/ESI/IT/TOF method described in this study for the first time was specific and sensitive with a LOD of 25 fmol in the MS mode and 500 fmol in the MS/MS mode for all compounds analyzed [53]. Moreover, the same LC/ESI/IT/TOF technique has been applied to identify and quantify, in the injured rat brain, new EC-like molecules (*N*-acylserines, *N*-acyldopamines, and *N*-acylglycines) potentially involved in a novel model of traumatic brain injury (TBI) (unpublished data). The recovery of NADA, NAGly, and NASer from tissues was 68.1 ± 1.3%, 49.1 ± 15.7%, and 42.1 ±15.9%. The LC/ESI/IT/TOF method described was again specific and sensitive with a LOD of 50 fmol in the MS mode and 1 pmol in the MS/MS mode for all compounds analyzed. The ratio between the [M + H]$^+$ peak areas of undeuterated (0.025–10 pmol) versus deuterated (1 pmol) NADA, NAGly, and NASer varied linearly with the amount of the respective deuterated standards. The quantification limit of compounds was 100 fmol and the reproducibility of the method was 95–99%. LC parameters were optimized to ensure good separation among the analytes, and an example of a chromatogram for the separation of *N*-acylglycines is shown in Figure 9.2A. This analysis provided high-resolution [M + H]$^+$ adducts and the quantification was performed by isotope dilution by using *m/z* values of 362.2696, 314.2702, and 340.2839, corresponding to the precursor ion of synthetic undeuterated NAGly, PalGly, and OlGly, respectively (Figure 9.2B).

The profiling of EC biosynthetic precursors could be also considered as part of endocannabinoidomics. The understanding of the regulation of EC levels under physio-pathological conditions, which is determined also by precursor availability, will be useful to facilitate the development of further drug candidates. In fact, very recently, a very sensitive and highly resolutive method to quantify NAPEs (*N*-acylphosphatidylethanolamines, the biosynthetic precursors of anandamide and other NAEs) and DAGs (the biosynthetic precursors of 2-AG and other 2-acylglycerols) in several tissues of mice fed with a dietary ω-3-PUFA supplementation [54] was set up, using the LC/MS/

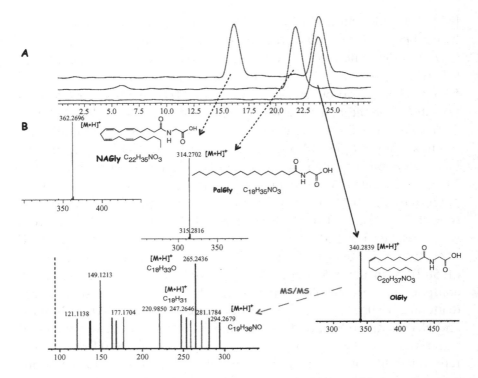

Fig. 9.2. Targeted profiling of N-acylglycines in lipid extracts using LC/MS/IT/TOF. (A) Representative extracted ion chromatogram of a prepurified brain lipid extract containing NAGly, PalGly, and OlGly. LC parameters were optimized to ensure good separation among the analytes. Shown in the panel (B) are the high-resolution MS spectra of the three compounds and the MS/MS analysis of OlGly showing the product ions with m/z 294.2679, 265.2436, and 247.2646, which in turn corresponds, respectively, to the $C_{19}H_{36}NO$, $C_{18}H_{33}O$, and $C_{18}H_{31}$ fragment.

IT/TOF technique described above. In particular, Figure 9.3 shows the total ion current for the separation of DAGs in a prepurified liver lipid extract and the corresponding MS and MS/MS spectra of the major component of the extract.

9.3.3 From Lipidomics of the Endocannabinoidome to Endocannabinoidomics

Since the complexity of the "EC metabolome" is increasing, it seems therefore evident that endocannabinoidomics includes not only the measurement of ECs and related mediators, but also their metabolic enzymes and their molecular targets. In this context, it is noteworthy a paper from Iannotti et al. [55], in which the authors report the complete

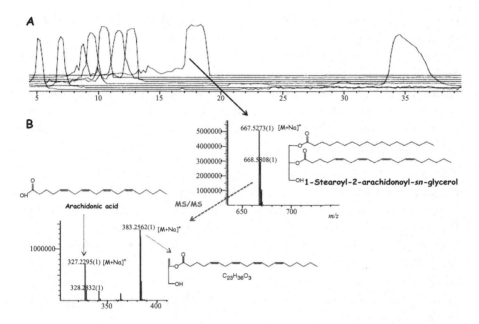

Fig. 9.3. (A) Representative extracted ion chromatogram of a prepurified liver lipid extract containing various DAG species (22:6–22:6, 18:2–22:6, 18:1–22:6, 16:0–22, 18:1–20:4, 16:0–20:4, in order of retention time, respectively). (B) Positive-ion electrospray mass spectrum of a major component of the extract, the 1-stearoyl-2-arachidonoyl-sn-glycerol precursor ion (m/z 667.5273) corresponding to the sodium adduct and the corresponding product ions for the CID of the fragment with m/z 667.5273 in the MS/MS spectra (m/z 327.2295 and 383.2562), which in turn correspond to the sodiated arachidonic acid and $C_{23}H_{36}O_3$ fragments, respectively. Adapted from Piscitelli et al. [54].

analysis of the "endocannabinoidome" in terms of EC and EC-related levels (by LC/MS), gene expression analysis (by RT-qPCR and western blot) and enzymatic assay (by radioactive assay) in both obese and lean Zucker rats, providing a new "high-throughput" analysis of the ECS.

9.4 NEW FRONTIERS

The introduction of matrix-assisted laser desorption/ionization (MALDI) imaging mass spectrometry (IMS) adds a new dimension to lipidomic studies; i.e., anatomical mapping of small molecules in the sample [56]. In fact, this technique is notably used to visualize various types of lipids, such as phospholipids, glycolipids, neutral lipids, and fatty acids [57,58], thus including precursors of ECs. Although MALDI-IMS is suitable for the analysis of major components in tissue sections, it has not yet been applied to detect minor constituents or molecules, which are not easy to ionize or are present only in trace amounts. However, research directed toward detecting and imaging small organic molecules has recently been expanding. Recently, MALDI-IMS has attracted great interest for the monitoring of drug delivery and metabolism, since it can visualize and distinguish the parent drug and its metabolites [59], thus opening a new frontier in pharmacology and toxicology.

No report exists concerning imaging analysis of the "endocannabinoidome" in tissue sections, possibly because ECs and EC-like molecules are present in biological matrices in very low concentrations. However, the use of TLC-Blot-MALDI-IMS, a technique in which lipids are first separated on TLC plates and then transferred to a polyvinylidene difluoride (PVDF) membrane followed by MALDI-IMS analysis [58], could increase the sensitivity and identify lipids at a low picomolar level, thus possibly also allowing the use of this technology for the detection *in situ* of EC-related molecules.

9.5 CONCLUSIONS

The development of analytical techniques for EC measurement in tissues and biological fluids has been helpful for understanding the physiological and pathological roles of these mediators in the brain as well as other organs and tissues [60]. In fact, it is also through the observation of how the levels of ECs change specifically in certain tissues, rather than others, that has made it possible to confirm the function of the ECS. Recent advances in liquid chromatography combined with tandem MS made possible not only the detection of increasingly lower concentrations of ECs in biological matrices, but also the discovery of novel bioactive fatty acid derivatives and their simultaneous monitoring together with the known members of a given subfamily of EC-like molecules.

In summary, increasing evidence confirms that the "endocannabinoidome," particularly if investigated using novel analytical approaches, is a new exciting field of biomedical research, which could be soon exploited to develop new therapeutic drugs and diagnostic biomarkers.

REFERENCES

[1] De Petrocellis L, Di Marzo V. An introduction to the endocannabinoid system: from the early to the latest concepts. Best Pract Res Clin Endocrinol Metab 2009;23(1):1–15.

[2] Bisogno T, Piscitelli F, Di Marzo V. Lipidomic methodologies applicable to the study of endocannabinoids and related compounds: endocannabinoidomics. Eur J Lipid Sci Technol 2009;111(1):53–63.

[3] Lichtman AH, Varvel SA, Martin BR. Endocannabinoids in cognition and dependence. Prostaglandins Leukot Essent Fatty Acids 2002;66(2–3):269–85.

[4] Pacher P, Batkai S, Kunos G. The endocannabinoid system as an emerging target of pharmacotherapy. Pharmacol Rev 2006;58(3):389–462.

[5] Di Marzo V. The endocannabinoid system in obesity and type 2 diabetes. Diabetologia 2008;51(8):1356–67.

[6] Maccarrone M. Endocannabinoids: friends and foes of reproduction. Prog Lipid Res 2009;48(6):344–54.

[7] Di Marzo V. "Endocannabinoids" and other fatty acid derivatives with cannabimimetic properties: biochemistry and possible physiopathological relevance. Biochim Biophys Acta 1998;1392(2–3):153–75.

[8] Hansen HS, Diep TA. N-Acylethanolamines, anandamide and food intake. Biochem Pharmacol 2009;78(6):553–60.

[9] Starowicz KM, Cristino L, Matias I, Capasso R, Racioppi A, Izzo AA, et al. Endocannabinoid dysregulation in the pancreas and adipose tissue of mice fed with a high-fat diet. Obesity (Silver Spring) 2008;16(3):553–65.

[10] Izzo AA, Piscitelli F, Capasso R, Marini P, Cristino L, Petrosino S, et al. Basal and fasting/refeeding-regulated tissue levels of endogenous PPAR-alpha ligands in Zucker rats. Obesity (Silver Spring) 2010;18(1):55–62.

[11] Ben-Shabat S, Fride E, Sheskin T, Tamiri T, Rhee MH, Vogel Z, et al. An entourage effect: inactive endogenous fatty acid glycerol esters enhance 2-arachidonoyl-glycerol cannabinoid activity. Eur J Pharmacol 1998;353(1):23–31.

[12] De Petrocellis L, Davis JB, Di Marzo V. Palmitoylethanolamide enhances anandamide stimulation of human vanilloid VR1 receptors. FEBS Lett 2001;506(3):253–6.

[13] Hansen HS. Palmitoylethanolamide and other anandamide congeners. Proposed role in the diseased brain. Exp Neurol 2010;224(1):48–55.

[14] Lambert DM, Di Marzo V. The palmitoylethanolamide and oleamide enigmas: are these two fatty acid amides cannabimimetic? Curr Med Chem 1999;6(8):757–73.

[15] Godlewski G, Offertaler L, Wagner JA, Kunos G. Receptors for acylethanolamides-GPR55 and GPR119. Prostaglandins Other Lipid Mediat. 2009;89(3–4):105–11.

[16] McKillop AM, Moran BM, Abdel-Wahab YH, Flatt PR. Evaluation of the insulin releasing and antihyperglycaemic activities of GPR55 lipid agonists using clonal beta-cells, isolated pancreatic islets and mice. Br J Pharmacol 2013;170(5):978–90.

[17] Mechoulam R, Ben-Shabat S, Hanus L, Ligumsky M, Kaminski NE, Schatz AR, et al. Identification of an endogenous 2-monoglyceride, present in canine gut, that binds to cannabinoid receptors. Biochem Pharmacol 1995;50(1):83–90.

[18] Sugiura T, Kondo S, Sukagawa A, Nakane S, Shinoda A, Itoh K, et al. 2-Arachidonoylglycerol: a possible endogenous cannabinoid receptor ligand in brain. Biochem Biophys Res Commun 1995;215(1):89–97.

[19] Ueda N, Tsuboi K. Discrimination between two endocannabinoids. Chem Biol 2012;19(5):545–7.

[20] Ueda N, Tsuboi K, Uyama T. Enzymological studies on the biosynthesis of *N*-acylethanolamines. Biochim Biophys Acta 2010;1801(12):1274–85.

[21] Ueda N, Tsuboi K, Uyama T, Ohnishi T. Biosynthesis and degradation of the endocannabinoid 2-arachidonoylglycerol. Biofactors 2011;37(1):1–7.

[22] Hansen KB, Rosenkilde MM, Knop FK, Wellner N, Diep TA, Rehfeld JF, et al. 2-Oleoyl glycerol is a GPR119 agonist and signals GLP-1 release in humans. J Clin Endocrinol Metab 2011;96(9):E1409–17.

[23] Bisogno T, Melck D, Bobrov M, Gretskaya NM, Bezuglov VV, De Petrocellis L, et al. *N*-Acyl-dopamines: novel synthetic CB(1) cannabinoid-receptor ligands and inhibitors of anandamide inactivation with cannabimimetic activity *in vitro* and *in vivo*. Biochem J 2000;351 Pt 3:817–24.

[24] Huang SM, Bisogno T, Trevisani M, Al-Hayani A, De Petrocellis L, Fezza F, et al. An endogenous capsaicin-like substance with high potency at recombinant and native vanilloid VR1 receptors. Proc Natl Acad Sci USA 2002;99(12):8400–5.

[25] Ross HR, Gilmore AJ, Connor M. Inhibition of human recombinant T-type calcium channels by the endocannabinoid *N*-arachidonoyl dopamine. Br J Pharmacol 2009;156(5):740–50.

[26] Starowicz K, Nigam S, Di Marzo V. Biochemistry and pharmacology of endovanilloids. Pharmacol Ther 2007;114(1):13–33.

[27] Burstein SH, Huang SM, Petros TJ, Rossetti RG, Walker JM, Zurier RB. Regulation of anandamide tissue levels by *N*-arachidonylglycine. Biochem Pharmacol 2002;64(7):1147–50.

[28] Kohno M, Hasegawa H, Inoue A, Muraoka M, Miyazaki T, Oka K, et al. Identification of *N*-arachidonylglycine as the endogenous ligand for orphan G-protein-coupled receptor GPR18. Biochem Biophys Res Commun 2006;347(3):827–32.

[29] Rimmerman N, Bradshaw HB, Hughes HV, Chen JS, Hu SS, McHugh D, et al. *N*-Palmitoyl glycine, a novel endogenous lipid that acts as a modulator of calcium influx and nitric oxide production in sensory neurons. Mol Pharmacol 2008;74(1):213–24.

[30] Bradshaw HB, Rimmerman N, Hu SS, Burstein S, Walker JM. Novel endogenous *N*-acyl glycines identification and characterization. Vitam Horm 2009;81:191–205.

[31] Kozak KR, Marnett LJ. Oxidative metabolism of endocannabinoids. Prostaglandins Leukot Essent Fatty Acids 2002;66(2–3):211–20.

[32] Woodward DF, Carling RW, Cornell CL, Fliri HG, Martos JL, Pettit SN, et al. The pharmacology and therapeutic relevance of endocannabinoid derived cyclo-oxygenase (COX)-2 products. Pharmacol Ther 2008;120(1):71–80.

[33] Verhoeckx KC, Voortman T, Balvers MG, Hendriks HF, Wortelboer HM, Witkamp RF. Presence, formation and putative biological activities of *N*-acyl serotonins, a novel class of fatty-acid derived mediators, in the intestinal tract. Biochim Biophys Acta 2011;1811(10):578–86.

[34] Bisogno T, Melck D, De Petrocellis L, Bobrov M, Gretskaya NM, Bezuglov VV, et al. Arachidonoylserotonin and other novel inhibitors of fatty acid amide hydrolase. Biochem Biophys Res Commun 1998;248(3):515–22.

[35] Maione S, De Petrocellis L, de Novellis V, Moriello AS, Petrosino S, Palazzo E, et al. Analgesic actions of *N*-arachidonoyl-serotonin, a fatty acid amide hydrolase inhibitor with antagonistic activity at vanilloid TRPV1 receptors. Br J Pharmacol 2007;150(6):766–81.

[36] Devane WA, Hanus L, Breuer A, Pertwee RG, Stevenson LA, Griffin G, et al. Isolation and structure of a brain constituent that binds to the cannabinoid receptor. Science 1992;258(5090):1946–9.

[37] Zoerner AA, Gutzki FM, Batkai S, May M, Rakers C, Engeli S, et al. Quantification of endocannabinoids in biological systems by chromatography and mass spectrometry: a comprehensive review from an analytical and biological perspective. Biochim Biophys Acta 2011;1811(11):706–23.

[38] Felder CC, Nielsen A, Briley EM, Palkovits M, Priller J, Axelrod J, et al. Isolation and measurement of the endogenous cannabinoid receptor agonist, anandamide, in brain and peripheral tissues of human and rat. FEBS Lett 1996;393(2–3):231–5.

[39] Giuffrida A, Rodriguez de Fonseca F, Piomelli D. Quantification of bioactive acylethanolamides in rat plasma by electrospray mass spectrometry. Anal Biochem 2000;280(1):87–93.

[40] Zoerner AA, Batkai S, Suchy MT, Gutzki FM, Engeli S, Jordan J, et al. Simultaneous UPLC-MS/MS quantification of the endocannabinoids 2-arachidonoyl glycerol (2AG), 1-arachidonoyl glycerol (1AG), and anandamide in human plasma: minimization of matrix-effects, 2AG/1AG isomerization and degradation by toluene solvent extraction. J Chromatogr B Analyt Technol Biomed Life Sci 2012;883–884:161–71.

[41] Kingsley PJ, Marnett LJ. Analysis of endocannabinoids by Ag+ coordination tandem mass spectrometry. Anal Biochem 2003;314(1):8–15.

[42] Matias I, Gonthier MP, Petrosino S, Docimo L, Capasso R, Hoareau L, et al. Role and regulation of acylethanolamides in energy balance: focus on adipocytes and beta-cells. Br J Pharmacol 2007;152(5):676–90.

[43] Stock K, Kumar J, Synowitz M, Petrosino S, Imperatore R, Smith ES, et al. Neural precursor cells induce cell death of high-grade astrocytomas through stimulation of TRPV1. Nat Med 2012;18(8):1232–8.

[44] Huang SM, Bisogno T, Petros TJ, Chang SY, Zavitsanos PA, Zipkin RE, et al. Identification of a new class of molecules, the arachidonyl amino acids, and characterization of one member that inhibits pain. J Biol Chem 2001;276(46):42639–44.

[45] Tan B, O'Dell DK, Yu YW, Monn MF, Hughes HV, Burstein S, et al. Identification of endogenous acyl amino acids based on a targeted lipidomics approach. J Lipid Res 2010;51(1): 112–9.

[46] Milman G, Maor Y, Abu-Lafi S, Horowitz M, Gallily R, Batkai S, et al. N-Arachidonoyl L-serine, an endocannabinoid-like brain constituent with vasodilatory properties. Proc Natl Acad Sci USA 2006;103(7):2428–33.

[47] Cohen-Yeshurun A, Trembovler V, Alexandrovich A, Ryberg E, Greasley PJ, Mechoulam R, et al. N-Arachidonoyl-L-serine is neuroprotective after traumatic brain injury by reducing apoptosis. J Cereb Blood Flow Metab 2011;31(8):1768–77.

[48] Han B, Wright R, Kirchhoff AM, Chester JA, Cooper BR, Davisson VJ, et al. Quantitative LC-MS/MS analysis of arachidonoyl amino acids in mouse brain with treatment of FAAH inhibitor. Anal Biochem 2013;432(2):74–81.

[49] Bystrowska B, Smaga I, Tyszka-Czochara M, Filip M. Troubleshooting in LC-MS/MS method for determining endocannabinoid and endocannabinoid-like molecules in rat brain structures applied to assessing the brain endocannabinoid/endovanilloid system significance. Toxicol Mech Methods 2014;24(4):315–22.

[50] Thieme U, Schelling G, Hauer D, Greif R, Dame T, Laubender RP, et al. Quantification of anandamide and 2-arachidonoylglycerol plasma levels to examine potential influences of tetrahydrocannabinol application on the endocannabinoid system in humans. Drug Test Anal 2014;6(1–2):17–23.

[51] Pavon FJ, Araos P, Pastor A, Calado M, Pedraz M, Campos-Cloute R, et al. Evaluation of plasma-free endocannabinoids and their congeners in abstinent cocaine addicts seeking outpatient treatment: impact of psychiatric co-morbidity. Addict Biol 2013;18(6):955–69.

[52] Siller M, Goyal S, Yoshimoto FK, Xiao Y, Wei S, Guengerich FP. Oxidation of endogenous N-arachidonoylserotonin by human cytochrome P450 2U1. J Biol Chem 2014;289(15):10476–87.

[53] Gatta L, Piscitelli F, Giordano C, Boccella S, Lichtman A, Maione S, et al. Discovery of prostamide F2alpha and its role in inflammatory pain and dorsal horn nociceptive neuron hyperexcitability. PLoS One 2012;7(2):e31111.

[54] Piscitelli F, Carta G, Bisogno T, Murru E, Cordeddu L, Berge K, et al. Effect of dietary krill oil supplementation on the endocannabinoidome of metabolically relevant tissues from high-fat-fed mice. Nutr Metab (Lond) 2011;8(1):51.

[55] Iannotti FA, Piscitelli F, Martella A, Mazzarella E, Allara M, Palmieri V, et al. Analysis of the "endocannabinoidome" in peripheral tissues of obese Zucker rats. Prostaglandins Leukot Essent Fatty Acids 2013;89(2–3):127–35.

[56] Caprioli RM, Farmer TB, Gile J. Molecular imaging of biological samples: localization of peptides and proteins using MALDI-TOF MS. Anal Chem 1997;69(23):4751–60.

[57] Fernandez JA, Ochoa B, Fresnedo O, Giralt MT, Rodriguez-Puertas R. Matrix-assisted laser desorption ionization imaging mass spectrometry in lipidomics. Anal Bioanal Chem 2011;401(1):29–51.

[58] Goto-Inoue N, Hayasaka T, Zaima N, Setou M. Imaging mass spectrometry for lipidomics. Biochim Biophys Acta 2011;1811(11):961–9.

[59] Sugiura Y, Setou M. Imaging mass spectrometry for visualization of drug and endogenous metabolite distribution: toward *in situ* pharmacometabolomes. J Neuroimmune Pharmacol 2010;5(1):31–43.

[60] Ligresti A, Petrosino S, Di Marzo V. From endocannabinoid profiling to 'endocannabinoid therapeutics'. Curr Opin Chem Biol 2009;13(3):321–31.

CHAPTER 10

Common Receptors for Endocannabinoid-Like Mediators and Plant Cannabinoids

Stephen P.H. Alexander

10.1 INTRODUCTION

The characteristic compounds found in the *Cannabis* plant are described elsewhere in this volume; in brief, these are lipophilic resorcinol metabolites, sometimes referred to as phytocannabinoids. Of these, Δ^9-tetrahydrocannabinol (THC) is easily the most renowned as it appears to be the major mood-altering substance in the *Cannabis* plant (for reviews, see Refs [1,2]). The development of radiolabeled cannabinoid derivatives allowed the identification of the CB_1 cannabinoid receptor, found to very

The Endocannabinoidome: The World of Endocannabinoids and Related Mediators. DOI: 10.1016/B978-0-12-420126-2.00010-9
Copyright © 2015 Elsevier Inc. All rights reserved

high levels in the central nervous system (CNS), and to lower densities in peripheral neural and other tissues. A second G protein-coupled receptor (GPCR) for cannabinoids, the CB_2 cannabinoid receptor, was identified initially outside the nervous system, primarily in immune-related tissues. Following identification of these receptors, further components of the endogenous signaling system were defined; the synthetic and transformative enzymes and the endocannabinoids (with the best characterized being anandamide and 2-arachidonoylglycerol, both arachidonic acid derivatives), which are described in more detail elsewhere in this book. In addition, relatively selective antagonists at CB_1 (rimonabant and AM251) and CB_2 (SR144528 and AM630) receptors have been described (see Ref. [3]).

In the last decade, it has become clear that further molecular targets of the endocannabinoid-like compounds and the plant-derived cannabinoids exist. This chapter will focus on evidence suggesting the potential for mediation of cannabinoid actions through CB_1, CB_2, and non-CB_1/CB_2 receptors. In particular, cannabinoid action through GPCR, ion channels, and nuclear hormone receptors (NHRs) is described (see Figure 10.1).

10.2 CB_1 AND CB_2 RECEPTORS

CB_1 and CB_2 receptors are, at least superficially, conventional members of the rhodopsin family of GPCRs. These are heptahelical polypeptide chains with extracellular amino termini and intracellular carboxy termini. The amino termini contain consensus sites for N-linked glycosylation, although the nature and significance of this post-translational modification has not been elucidated. The binding site for ligands appears to be in the plane of the plasma membrane, which is common to all rhodopsin type, family A GPCR. However, less widespread among the family is the unusual route to this binding site that ligands appear to take. Thus, for both CB_1 [4] and CB_2 [5] receptors, molecular dynamics studies suggest that ligands enter in a "sideways mode," making use of an entry site via the lipid bilayer. The best current hypothesis has the ligands "dissolving" in the plasma membrane prior to delivery to the receptor-binding site following lateral diffusion.

Within an individual species, comparing CB_1 and CB_2 receptors shows limited identity; in man, for example, there is only 44% amino

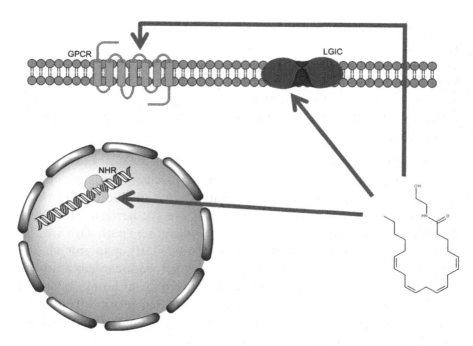

Fig. 10.1. Multiple cellular targets for anandamide. Anandamide is the first ligand identified to act at members of three receptor superfamilies: GPCRs, LGIC, and NHRs. Anandamide generated inside the cell is able to act in the nucleus at PPARs, on the intracellular face of the plasma membrane at TRPV1 receptors, and at GPCR either from the external face of the plasma membrane (as illustrated here) or by "sideways" access through the plane of the plasma membrane.

acid homology between the two receptors [6]. Ensembl (www.ensembl.org) suggests that CB_1 receptors are well conserved across mammalian species: 97–100% for gorilla, chimpanzee, orangutan, cow, pig, sheep, guinea pig, rat, hamster, and mouse. The similarity is less in birds and fish: 93% for duck, 89% for chicken, and 70% for zebrafish. For CB_2 receptors, sequence similarity is high in primates (100% for chimpanzee, 99% for gorilla, and 98% for orangutan). However, this is reduced in other mammals: sheep and mouse (83%), cow and pig (82%), guinea pig (75%), and rat (70%), and even lower in birds and fish: duck (51%), chicken (50%), and zebrafish (39%).

10.2.1 Agonist-Evoked Signaling

The CB_1 receptor is a GPCR, which couples primarily to the $G_{i/o}$ family of G proteins, leading to inhibition of adenylyl cyclase activity. In addition, agonist activation leads to the opening of potassium channels and closing of calcium channels, thought to be mediated by the G protein. This regulation of ion channel function, in particular, is associated with an inhibition of neurotransmitter release in the CNS, where GABA and glutamate release are prominent targets. The CB_2 cannabinoid receptor is also $G_{i/o}$-coupled, but appears to couple solely to inhibition of adenylyl cyclase activity, without the concomitant regulation of ion channels.

The best candidates for endogenous ligands of the CB_1 and CB_2 cannabinoid receptors are arachidonate derivatives, eicosanoids, which may be divided into amide (anandamide) and ester (2-arachidonoylglycerol, 2AG) analogs. These two lipid derivatives are synthesized and hydrolyzed by distinct enzymatic pathways, although these pathways are also involved in turnover of analogs with distinct fatty acid side-chains. These analogs have sometimes been termed endocannabinoid-like molecules because of their structural similarities, although relatively few have been shown to activate CB_1 or CB_2 receptors. Intriguingly, the two endogenous cannabinoid agents have distinct affinity/efficacy profiles. Anandamide behaves as a partial agonist at both CB_1 and CB_2 receptors, and also appears to be less potent than the full agonist 2AG [2]. In the brain, tissue levels of 2AG are in great excess compared to anandamide [7–9], which might be interpreted as 2AG being the better candidate for the major endogenous cannabinoid. However, tissue levels of these endocannabinoids may not necessarily reflect extracellular levels [10], and it is entirely possible that the interplay between the two endogenous agonists with distinct efficacies has some impact on signaling.

The major psychoactive ingredient in extracts from the *Cannabis* plant, THC, is also a partial agonist at CB_1 and CB_2 receptors, although it is unclear what significance this may have for the widespread consumption of cannabis. In recombinant systems, THC is slightly CB_2-preferring (less than 10-fold), but appears to be about 10-fold more potent than anandamide in binding and functional assays [11]. Although the other constituents of the *Cannabis* plant have not been exhaustively assayed for CB_1/CB_2 receptor effects, none so far show the same level of interaction as THC, indicating that other molecular targets are involved in the actions of cannabidiol, tetrahydrocannabivarin, etc.

10.2.2 Constitutive Activity and Inverse Agonism

The phenomenon of constitutive activity, whereby GPCR appear to be at an active state in the absence of added agonist, is a common observation in recombinant systems. This has been interpreted as the receptor fluctuating between different activation states (known as R and R*) and overexpression prompting activity of the downstream signaling pathway possibly through a simple increase in the probability of the R* form interacting with the G protein. In parallel with these observations came the concept of inverse agonism, whereby ligands could bind to the receptor and revert it to a baseline state (from R* to R), reducing the activation of the downstream cascade. Notably, the two archetypal antagonists at CB_1 and CB_2 receptors have both been described to have inverse agonist properties. Thus, rimonabant has been reported to have inverse agonist effects at CB_1 receptors expressed in brain and other organs, as well as in cells [12–15]. This effect may be concentration-dependent in that nanomolar concentrations of rimonabant appear to act as a "simple" competitive antagonist, but micromolar concentrations act as an inverse agonist [16]. SR 144528 was the first CB_2 receptor-selective antagonist described, and has also been described to act as an inverse agonist in recombinant systems [17,18]. As noted recently, however, the difficulty in establishing an absence of endogenous cannabinoid tone in contributing to apparent constitutive activity of cannabinoid receptors means that it may be unsafe to ascribe effects of cannabinoid receptor antagonists in the absence of added agonist to inverse agonism [19].

10.2.3 CB Cannabinoid Receptor Function *In Vitro* and *In Vivo*

Given the coupling of CB_1 receptors to the pertussis toxin-sensitive $G_{i/o}$ family of G proteins, the most common assays of CB_1 receptor function involve enhancement of $[^{35}S]$-GTPγS binding to preparations from brain tissue or cells [20] and the inhibition of cyclic AMP accumulation or activation of extracellular signal-regulated protein kinase (ERK) phosphorylation in cell preparations [21]. The more traditional pharmacological approaches also allow an assessment of CB_1 receptor function using isolated tissue preparations, such as the electrically evoked twitch responses in the guinea-pig ileum or mouse vas deferens. CB_1 receptor activation results in decreased transmitter release from the parasympathetic [22] and sympathetic [23] nerve terminals, respectively, leading to reduced contractile responses.

Analysis of CB_2 receptor function in cells utilizes similar assays: [^{35}S]-GTPγS binding [24] and ERK phosphorylation [25]. Due to the association of CB_2 receptors with cells of the immune system, there are no comparable tissue-based assays of CB_2 receptor function, but cultured macrophages appear to be a useable cellular assay for native CB_2 receptor coupling to anti-inflammatory processes [26].

Cannabinoid administration to rodents *in vivo* results in the "classical tetrad" of catalepsy, hypolocomotion (hypokinesia), antinociception, and hypothermia, in a manner sensitive to rimonabant [27]. In parallel with many observations of the effects of cannabis preparations in man, cannabinoid administration in animals evokes changes in cognition, memory, anxiety, intraocular pressure, appetite, and emesis, as well as neuronal survival, cancer proliferation, inflammation, and immune responses [28]. These effects appear to be largely or completely due to activation of either CB_1 and/or CB_2 cannabinoid receptors, as they are reversed by the selective antagonists/inverse agonists described above.

10.3 BEYOND CB_1 AND CB_2 RECEPTORS: CANNABINOID RECEPTOR-LIKE GPCR

As indicated above, endocannabinoids acting at the well characterized CB_1 and CB_2 cannabinoid receptors are 2-arachidonoylglycerol and anandamide. In contrast, *N*-arachidonoylglycine (NAGly), a potential oxidative metabolite of anandamide [29], is ineffective at CB_1 or CB_2 receptors [30]. NAGly has been observed to have antinociceptive effects *in vivo* [30,31], but the molecular locus of this action has yet to be identified, with candidates including glycine transporter GlyT2 [32], glycine receptors containing α2 and α3 subunits [33], and calcium-activated potassium channels [34,35]. However, NAGly was also reported to activate recombinant GPR18, an orphan GPCR, at nanomolar concentrations [36]. More recently, this effect of NAGly was confirmed with both recombinant and native GPR18 expressed in a human endometrial cell line [37]. Intriguingly, this same study identified some further interesting pharmacology of this receptor. Alongside NAGly, anandamide and THC also acted as micromolar potency, full agonists at this receptor. Cannabidiol and AM251 appeared to act as partial agonists at this receptor with EC_{50} values in excess of 50 μM [37], while rimonabant and

SR 144528 appeared ineffective. There remains an unanswered question about the possible contribution of GPR18 to the effects of *Cannabis* consumption in man.

GPR55 has also been described as a cannabinoid receptor-like receptor. Like GPR18, GPR55 has only limited structural similarity to CB_1 or CB_2 cannabinoid receptors [38]. Although it has a variable pharmacological profile, termed "enigmatic" by one author [39], a relatively consistent property of GPR55 is that it responds to AM251, a CB_1 cannabinoid receptor antagonist/inverse agonist structurally related to rimonabant [40–42]. The initial description also suggested that a number of endocannabinoid-like molecules acted as agonists, in particular, anandamide, OEA, PEA, and 2AG [42]. Notably, THC also appeared to be a potent agonist, while cannabidiol appeared to act as an antagonist. In contrast, a separate study failed to observe agonist effects of any of the endocannabinoid-like molecules identified above or THC [43]. Currently, the best candidate for an endogenous agonist at GPR55 is a lysophospholipid derivative of 2AG, 2-arachidonoylglycerolphosphoinositol [44]. Although antagonists of GPR55 have been described [45,46], they have not yet been widely applied, so the physiological significance of GPR55 as a potential mediator of effects of cannabinoids is not yet defined.

A third cannabinoid receptor-like GPCR is GPR119. GPR119 has been proposed to be a receptor for *N*-oleoylethanolamine [47] and 2-oleoylglycerol [48], analogs of anandamide and 2AG, respectively (see elsewhere in this volume). Longer chain, polyunsaturated analogs appeared ineffective as agonists, differentiating GPR119 from the conventional CB_1 and CB_2 cannabinoid receptors. A number of synthetic analogs with activity at GPR119 has been described, allowing a proposed role for GPR119 in satiety and regulation of metabolism.

As with many other GPCR, there is evidence for CB_1 receptors forming multimers. The best described of these is a complex of dopamine D_2 and CB_1 receptors [49–51]. Further candidate partners for heteromultimerization include opioid receptors [52–54], OX_1 orexin receptors [55], A_{2A} adenosine [56], and $β_2$-adrenoceptors [57]. These heteromultimers are likely to have complex pharmacological profiles, as well as complicated signaling sequelae. Whether CB_1 or CB_2 receptors heteromultimerize with the cannabinoid receptor-like GPCR (GPR18, GPR55, and GPR119; see below) appears not yet to have been investigated.

10.4 LIGAND- AND VOLTAGE-GATED ION CHANNELS

TRPV1 receptors remain the best candidates for a classical ligand-gated ion channel (LGIC) target for endocannabinoids, particularly anandamide [58] (see Figure 10.1). In contrast, the quintessential TRPV1 ligand capsaicin does not bind to CB_1 receptors [59]. There is good evidence for a number of members of the transient receptor potential family to be regulated by cannabinoids (see Ref. [60]). In particular, CBD is an agonist at TRPV1 [61], TRPV2 [62], and TRPV3 [63], while TRPA1 and TRPM8 respond to THC and CBD [64]. The TRPs are often described as nonselective cation channels, although the most common method to measure their function is based on the measurement of intracellular calcium ions. Given their predominant (although not exclusive) location in sensory neurons, the anticipated impact of cannabinoids is to act as pronociceptive agents, in contrast to the general antinociceptive properties mediated through CB_1 and CB_2 cannabinoid receptors.

Direct effects of endocannabinoids have been reported on many other transmitter-gated channels, particularly the Cys-loop family. Thus, endocannabinoids reportedly inhibit glycine receptors [65], $GABA_A$ receptors [65,66], α7-nicotinic receptors [67], and 5-HT_3 receptors [68,69]. In contrast, NMDA glutamate receptors are reported to be positively modulated directly by anandamide [70]. In the majority of cases, the potential relevance of these effects to cannabinoid action *in vivo* is unclear.

Direct effects of cannabinoids on cation channels have been suggested. Both THC and CBD inhibited recombinant T-type (Ca_v3) voltage-sensitive calcium channels at micromolar concentrations [71]. Similarly, anandamide has been reported to inhibit T-type channels, and this has been claimed to contribute to analgesic effects of cannabinoids and related lipoamino acids [72–74].

10.5 NUCLEAR HORMONE RECEPTORS

As illustrated above, cannabinoids are active at members of the GPCR superfamily, as well as examples of the transmitter-gated channel receptor family. Given the hydrophobic nature of the cannabinoids, it should come as no surprise that there is the potential for signaling events evoked by cannabinoids inside the cell, particularly through regulation of NHRs (see Figure 10.1). To date, a significant focus of non-CB_1, non-CB_2 receptor-mediated effects of cannabinoids has been the peroxisome proliferator-activated receptors, PPARs (for earlier reviews, see Refs [75,76]). PPARs are widely distributed, but are prominent in a number of tissues, dependent on the subtype. Thus, PPARα is found in the liver, skeletal muscle, kidney, and heart, while PPARγ is particularly well expressed in adipose tissue, lymphoid tissues, and the large intestine.

10.5.1 The Classical View of PPARs

The traditional view of PPARs as NHRs is that they act in the nucleus as hormone receptors (for review, see Ref. [77]). That is, upon ligand activation, they regulate the transcription of DNA into messenger RNA. The archetypal ligands for PPARs are medium- and long-chain fatty acids, usually with various degrees of unsaturation. These circulate in the blood plasma bound to albumin. Through the action of members of the solute carrier family of transporters, the FATP/SLC27A fatty acid transport proteins, they accumulate in cells bound to members of the fatty acid-binding protein (FABP) family. The FABPs are able to shuttle fatty acids around the cell for metabolism, relaying fatty acids, as CoA esters, to members of the ABCD family of peroxisomal transporters, and the subsequent β-oxidation inside the peroxisome. Alternatively, FABPs are able to deliver long chain polyunsaturated fatty acids, such as arachidonic acid, for oxidative metabolism through cyclooxygenases and lipoxygenases. A third alternative is that FABPs can relay their fatty acid cargo to PPARs in the nucleus. In contrast to the classical view, based on the steroid hormone (Class I nuclear hormone) receptors, which in the absence of ligand are normally resident in the cytosol bound to chaperone proteins such as hsp90, PPARs are predominantly nuclear-located for their life cycle. A further distinction from the steroid hormone receptors is that they are obligate heterodimers, requiring retinoid X receptors (RXR) as partners to allow functional activity. This fact implies a further complication of the signaling process, since PPAR-mediated signaling not only requires the presence of an appropriate concentration of PPAR ligand, but also the simultaneous presence of an appropriate concentration of an RXR ligand, 9-*cis*-retinoic acid. Under normal nutrition, 9-*cis*-retinoic acid, derived from vitamin A, is considered to be at saturating levels for RXR activity, meaning that the activity of the PPAR:RXR heterodimer is primarily dependent on the variation in levels of PPAR agonists. Further modulation of PPAR signaling comes from the involvement of additional protein partners, corepressors and coactivators. Corepressors of PPAR signaling, such as NCoR1, recruit histone deacetylases to increase DNA binding to histones, hence impeding gene transcripton. Corepressors are usually described as being prebound to the PPAR, but are then displaced upon agonist binding by coactivators, such as NCoA1. The PPAR complex in the nucleus is bound to peroxisome proliferator response elements (PPRE). PPRE are sequences of DNA in the promoter region of particular genes, which present as repeats of nucleotides, e.g., AGGTCAnAGGTCA. These consensus sequences are typically found upstream of genes coding for proteins involved in lipid metabolism, such as apolipoproteins and lipoprotein lipase. Intriguingly, PPRE are also found upstream of many of the proteins described in the fatty acid life cycle described above. In particular, members of the fatty acid transport proteins and FABP families are upregulated by PPARγ activation, as well as PPARγ itself [77].

10.5.2 The Pharmacology of PPARs

The conventional view of PPARs is that there is no single endogenous ligand, which activates them, but more likely they act as local, generalized lipid sensors. A number of endogenous activators have been described for PPARs, including arachidonic acid, 13-hydroxyoctadecanoic acid, oleic acid, and leukotriene B_4.

Of the three forms of PPARs, two are currently exploited for therapeutic benefit. PPARα is targeted in hyperlipidemia using a class of compounds termed fibrates, exemplified by gemfibrozil and fenofibrate. PPARγ is targeted in type II diabetes, using agents collectively termed glitazones, typified by the presence of a thiazolidinedione moiety, such as pioglitazone, rosiglitazone, and troglitazone. Apparently paradoxically, PPARγ agonists are associated with weight gain, which seems at odds with a primary goal of the treatment of type II diabetes, which is to reduce patient weight. This is explained as a crucial role for PPARγ, particularly the PPARγ2 splice variant, in the maturation of adipocytes from preadipocytes. This differentiation step allows recovery of insulin sensitivity, but consequently allows greater accumulation of lipid in adipose tissue.

Selective agonists for PPARβ have been described, such as GW501516 and GW0742. Although such ligands have been suggested to have potential in the therapy of lipid disorders, as yet there are no licensed medicines targeting this receptor.

Selective antagonists have also been generated for each subtype of PPAR. Thus, GW9662 [78] and T0070907 [79] are nanomolar affinity antagonists at PPARγ. GW6471 [80] and GSK0660 [81] are selective antagonists at PPARα and PPARβ, respectively, with submicromolar affinity. MK886 also blocks PPARα with submicromolar affinity [82], but in addition reduces 5-lipoxygenase activity through inhibition of the associated activating protein [83].

Alongside the metabolic effects of PPAR action, there are other intriguing effects of PPAR activation, which are more recently identified. These include therapeutic potential identified in animal models of nicotine reward and relapse [84], Alzheimer's disease [85,86], Huntington's disease [87,88], Parkinson's disease [89], and epilepsy [90,91]. *In vitro* PPARγ activation leads to survival of cerebral cortical neurons [92].

10.5.3 PPAR Activation by Phytocannabinoids and Endocannabinoids

The most frequent method to assess PPAR activation is the use of reporter gene assays, whereby a readily measurable gene product, such as luciferase or alkaline phosphatase, is combined with a PPRE-expressing promoter. Necessarily, this method requires the transfection of intact, metabolically active cells in combination with extended incubation periods, which means that there is a distinct possibility that ligands, particularly endogenous lipid derivatives such as the endocannabinoids, may be metabolized prior to encountering the PPARs in the nucleus. Indeed, there is good evidence that this is the case (see below). Thus, reporter gene assays require additional validation, such as the use of cell-free assays for ligand occupancy measurements. This is most often conducted using truncated versions of the PPARs, particularly the ligand-binding domain, in combination with fluorescent ligands.

Using cellular assays of PPAR activity, anandamide has been described to activate both PPARα and PPARγ [93,94], while THC [95] and CBD [96] appear to show some selectivity for PPARγ. In addition,

OEA [94,97] and PEA [98], inactive at CB_1/CB_2 receptors, showed reasonable potency at PPARα. The picture is incomplete, however, since not all combinations of cannabinoids and PPARs have been reported.

At a cellular level, cannabinoids have been shown to activate PPARs expressed endogenously. For example, THC (and pioglitazone) evoked neuroprotection over 48 h in SH-SY5Y cells, in a manner sensitive to the PPARγ antagonist T0070907, but not the CB_1 receptor antagonist AM251 [99]. A potential indirect effect of cannabinoids on PPARγ has been suggested. Thus, cannabidiol-induced apoptosis of human lung cancer cells *in vitro* and upregulated COX-2 and PPARγ in these cells [100].

THC-evoked relaxation of rat precontracted aorta could be blocked by a PPARγ antagonist [95,101], as could that evoked by CBD [96] and anandamide [102]. In rat hippocampal organotypic cultures, PEA was able to reduce β-amyloid-evoked astrogliosis in a manner inhibited by PPARα, but not PPARγ, antagonists [103]. PEA, in combination with the FAAH inhibitor URB597, enhanced trabecular meshwork aqueous humor outflow in porcine eyes *in vitro* [104]. This was not blocked by rimonabant, but was partially reversed by SR144528 or GW6471, a PPARα antagonist. PEA-evoked effects on aqueous humor outflow were blocked by PD98059, which also blocked the PEA-evoked elevation in ERK1/2 phosphorylation.

The use of selective antagonists has substantiated evidence for cannabinoids activating PPARs *in vivo*. For example, PEA administration evoking antinociceptive effects in a mouse model of neuropathic pain was sensitive, at least in part, to coadministration of a PPARγ antagonist [105]. PEA administered directly to the periaqueductal gray also exerted antinociceptive effects. However, the effects of PEA administration on neuronal activity in this nucleus were blocked by a PPARα antagonist [106]. Intriguingly, AM251, the CB_1 receptor antagonist/inverse agonist and GPR55 agonist enhanced the effects of intra-PAG PEA.

Intrahippocampal Aβ infusion leading to spatial impairment and local inflammatory mediator (TNFα, active caspase 3, NFκB) and TUNEL-positive neurons could be diminished by the synthetic cannabinoid WIN55212-2 [107]. This effect could be partially reversed by CB_1, CB_2, or PPARγ antagonists, and completely reversed by a combination of these three agents. The FAAH inhibitor URB597 had no effect on cocaine or morphine-induced effects in the shell of the nucleus accumbens [108]. However, it did block the effects of nicotine in a manner, which could be reversed by rimonabant or MK886 suggesting a joint role of CB_1 and PPARα in these effects. In a model of cancer, A549-xenografted nude mice treated with CBD showed regression of the tumor, which was sensitive to GW9662 [100].

Although a number of groups have identified that FABPs may bind and deliver endocannabinoids and endocannabinoid-like molecules to PPARs and for metabolism [109–111], a full investigation of the role of FABPs in the life cycle of these mediators is lacking.

An alternative approach to antagonist studies is to make use of models of selective gene disruption. Thus, PPARα-knockout mice are viable [112], as are mice in which the gene encoding PPARβ is disrupted [113,114]. However, embryonic lethality is a consequence of universal disruption of PAARγ [115–117], which necessitated the development of conditional PPARγ-knockout mice [118], which should allow further investigations of PPARγ responses *in vivo*.

PPARα-knockout mice have been used repeatedly in studies of endocannabinoids and endocannabinoid-like molecules. These mice lose antinociceptive responses to PEA, and the selective PPARα agonist GW7647 [119–121]. Further, OEA effects on feeding behaviors [94], lipolysis [97], and neuroprotection [122] were lost when PPARα was knocked out. In contrast, effects of OEA on visceral pain [123] and intestinal motility [124] were maintained in PPARα-knockout mice. As yet, however, there are no descriptions of the effects of PPARβ or PPARγ gene disruption on responses to endocannabinoids or endocannabinoid-like responses. Perhaps of more significance is the lack of data on the effects of phytocannabinoids in animals in which any of the genes encoding PPARs are disrupted.

10.6 AMPLIFICATION OF ENDOCANNABINOID TONE

The initial rationale for the use of inhibitors of endocannabinoid hydrolysis and transport was based around amplifying the endogenous tone at CB_1/CB_2 receptors. While this is clearly the case in many experimental approaches, there is also evidence for the indirect activation of PPARs through the use of these inhibitors. Thus, MK886, but not rimonabant, prevented the effects of the FAAH inhibitor URB597 on nicotine activation of mesolimbic dopaminergic neurons *in vitro*; OEA and PEA mimicked these effects, in a manner also sensitive to MK886 [125]. Intraplantar administration of URB597 increased tissue levels of anandamide and 2AG (without significantly altering OEA or PEA), and allowed a reduction in pain behaviors, which were sensitive to local administration of a PPARα, but not a PPARγ antagonist [126]. Intraplantar URB597 administration also inhibited carrageenan-induced expansion of peripheral receptive field in spinal recordings in a manner that was mimicked by a selective PPARα agonist and which could be prevented by a selective PPARα antagonist [127]. The putative endocannabinoid transport inhibitors AM404 and UCM707 have been used to amplify endocannabinoid tone; the effects of which appear to be mediated by CB_1, CB_2, and PPARγ both *in vitro* [128] and *in vivo* [129].

Although FAAH hydrolyses a variety of *N*-acylethanolamines, an alternative metabolic route for hydrolysis of medium chain length, saturated *N*-acylethanolamines, such as PEA has been suggested. A selective *N*-acylethanolamine acid amidase (NAAA) inhibitor has been reported to elevate levels of PEA, but not anandamide or 2AG, in cultures of rat dorsal root ganglion neurons [130]. This inhibitor, ARN077, mimicked the inhibitory effect of exogenous PEA administration on depolarization-evoked calcium elevations in these cells; an effect blocked by the PPARα antagonist GW6471. Topical administration of ARN077 *in vivo* exhibited antinociceptive effects in two mouse models of pain behaviors; effects prevented by coadministration of GW6471 or in *ppara*-null mice [131].

Collectively, these data might be taken to indicate that PPARs are only activated significantly *in vivo* by endocannabinoid-like molecules when levels are pharmacologically amplified. The interpretation of these data is further complicated in that the products of oxidative metabolism of endocannabinoids are also PPAR agonists, including a 15-lipoxygenase product of 2AG, active at PPARα [132], an epoxygenase product of 2AG, also active at PPARα [133], and a cyclooxygenase-2 product of 2AG, active at PPARβ [134].

Thus, although the use of PPAR antagonists or models in which genes encoding PPARs are selectively disrupted identifies the PPARs in the mediation of responses to exogenous and endogenous cannabinoids, there remains the possibility of some doubt about whether cannabinoids are really PPAR ligands in intact tissue preparations or *in vivo*. Overall, endocannabinoid potencies as PPAR agonists are relatively low compared to their potencies as agonists of canonical CB_1/CB_2 cannabinoid receptors, which might be taken as evidence that endocannabinoids are poor candidates as PPAR ligands *in vivo*. However, a further influence is the level of background PPAR agonist tone, which may be up to 20 μM for intracellular long chain fatty acids [135]. These concentrations are sufficient to occupy PPARs in cell-free systems, and so while this background level will vary depending on the cell type and the active state of the cell, fluctuations in levels of intracellular endocannabinoid and endocannabinoid-like molecules may well prove sufficient to activate PPARs *in vivo*.

10.7 CONCLUDING REMARKS

A final point to make is that the marked similarities observed in functional effects of cannabinoid GPCR and PPARs in animal models assessing pain, inflammation, neuronal survival, feeding behaviors, and lipid turnover highlights the potential benefit of dual activation of these signaling pathways.

REFERENCES

[1] Mechoulam R, Parker LA. The endocannabinoid system and the brain. Annu Rev Psychol 2013;64:21–47.

[2] Pertwee RG, Howlett AC, Abood ME, Alexander SPH, Di Marzo V, Elphick MR, et al. International Union of Basic and Clinical Pharmacology. LXXIX. Cannabinoid receptors and their ligands: beyond CB_1 and CB_2. Pharmacol Rev 2010;62:588–631.

[3] Alexander SPH, Benson HE, Faccenda E, Pawson AJ, Sharman JL, Spedding M, et al. The concise guide to pharmacology 2013/14: G protein-coupled receptors. Br J Pharmacol 2013;170:1459–581.

[4] Lynch DL, Reggio PH. Cannabinoid CB1 receptor recognition of endocannabinoids via the lipid bilayer: molecular dynamics simulations of CB1 transmembrane helix 6 and anandamide in a phospholipid bilayer. J Comput Aided Mol Des 2006;20:495–509.

[5] Pei Y, Mercier RW, Anday JK, Thakur GA, Zvonok AM, Hurst D, et al. Ligand-binding architecture of human CB_2 cannabinoid receptor: evidence for receptor subtype-specific binding motif and modeling GPCR activation. Chem Biol 2008;15:1207–19.

[6] Munro S, Thomas KL, Abu-Shaar M. Molecular characterization of a peripheral receptor for cannabinoids. Nature 1993;365:61–5.

[7] Richardson D, Ortori CA, Chapman V, Kendall DA, Barrett DA. Quantitative profiling of endocannabinoids and related compounds in rat brain using liquid chromatography-tandem electrospray ionization mass spectrometry. Anal Biochem 2007;360:216–26.

[8] Bradshaw HB, Rimmerman N, Krey JF, Walker JM. Sex and hormonal cycle differences in rat brain levels of pain-related cannabimimetic lipid mediators. Am J Physiol Regul Integr Comp Physiol 2006;291:R349–58.

[9] Valenti M, Vigano D, Casico MG, Rubino T, Steardo L, Parolaro D, et al. Differential diurnal variations of anandamide and 2-arachidonoyl-glycerol levels in rat brain. Cell Mol Life Sci 2004;61:945–50.

[10] Sarmad S, Alexander SPH, Barrett DA, Marsden CA, Kendall DA. Depolarizing and calcium-mobilizing stimuli fail to enhance synthesis and release of endocannabinoids from rat brain cerebral cortex slices. J Neurochem 2011;117:665–77.

[11] Felder CC, Joyce KE, Briley EM, Mansouri J, Mackie K, Blond O, et al. Comparison of the pharmacology and signal transduction of the human cannabinoid CB_1 and CB_2 receptors. Mol Pharmacol 1995;48:443–50.

[12] Meschler JP, Kraichely DM, Wilken GH, Howlett AC. Inverse agonist properties of N-(piperidin-1-yl)-5-(4-chlorophenyl)-1-(2, 4-dichlorophenyl)-4-methyl-1H-pyrazole-3-carboxamide HCl (SR 141716A) and 1-(2-chlorophenyl)-4-cyano-5-(4-methoxyphenyl)-1H-pyrazole-3-carboxylic acid phenylamide (CP-272871) for the CB(1) cannabinoid receptor. Biochem Pharmacol 2000;60:1315–23.

[13] MaClennan SJ, Reynen PH, Kwan J, Bonhaus DW. Evidence for inverse agonism of SR 141716A at human recombinant cannabinoid CB_1 and CB_2 receptors. Br J Pharmacol 1998;124:619–22.

[14] Pan XH, Ikeda SR, Lewis DL. SR 141716A acts as an inverse agonist to increase neuronal voltage-dependent Ca^{2+} currents by reversal of tonic CB1 cannabinoid receptor activity. Mol Pharmacol 1998;54:1064–72.

[15] Landsman RS, Burkey TH, Consroe P, Roeske WR, Yamamura HI. SR 141716A is an inverse agonist at the human cannabinoid CB1 receptor. Eur J Pharmacol 1997;334:R1–2.

[16] Sim-Selley LJ, Brunk LK, Selley DE. Inhibitory effects of SR 141716A on G-protein activation in rat brain. Eur J Pharmacol 2001;414:135–43.

[17] Shire D, Calandra B, Bouaboula M, Barth F, Rinaldi-Carmona M, Casellas P, et al. Cannabinoid receptor interactions with the antagonists SR 141716A and SR 144528. Life Sci 1999;65:627–35.

[18] Portier M, Rinaldi-Carmona M, Pecceu F, Combes T, Poinot-Chazel C, Calandra B, et al. SR 144528, an antagonist for the peripheral cannabinoid receptor that behaves as an inverse agonist. J Pharmacol Exp Ther 1999;288:582–9.

[19] Howlett AC, Reggio PH, Childers SR, Hampson RE, Ulloa NM, Deutsch DG. Endocannabinoid tone versus constitutive activity of cannabinoid receptors. Br J Pharmacol 2011;163:1329–43.

[20] Griffin G, Atkinson PJ, Showalter VM, Martin BR, Abood ME. Evaluation of cannabinoid receptor agonists and antagonists using the guanosine-5'-O-(3-[^{35}S]thio)-triphosphate binding assay in rat cerebellar membranes. J Pharmacol Exp Ther 1998;285:553–60.

[21] Bouaboula M, Poinot-Chazel C, Bourrie B, Canat X, Calandra B, Rinaldi-Carmona M, et al. Activation of mitogen-activated protein kinases by stimulation of the central cannabinoid receptor CB1. Biochem J 1995;312:637–41.

[22] Pertwee RG, Fernando SR, Nash JE, Coutts AA. Further evidence for the presence of cannabinoid CB_1 receptors in guinea-pig small intestine. Br J Pharmacol 1996;118:2199–205.

[23] Devane WA, Hanus L, Breuer A, Pertwee RG, Stevenson LA, Griffin G, et al. Isolation and structure of a brain constituent that binds to the cannabinoid receptor. Science 1992;258:1946–9.

[24] Griffin G, Wray EJ, Tao Q, McAllister SD, Rorrer WK, Aung MM, et al. Evaluation of the cannabinoid CB2 receptor-selective antagonist, SR144528: further evidence for cannabinoid CB2 receptor absence in the rat central nervous system. Eur J Pharmacol 1999;377:117–25.

[25] Bouaboula M, Poinot-Chazel C, Marchand J, Canat X, Bourrie B, Rinaldi-Carmona M, et al. Signaling pathway associated with stimulation of CB2 peripheral cannabinoid receptor. Involvement of both mitogen-activated protein kinase and induction of Krox-24 expression. Eur J Biochem 1996;237:704–11.

[26] Ross RA, Brockie HC, Pertwee RG. Inhibition of nitric oxide production in RAW264.7 macrophages by cannabinoids and palmitoylethanolamide. Eur J Pharmacol 2000;401:121–30.

[27] Fox A, Kesingland A, Gentry C, McNair K, Patel S, Urban L, et al. The role of central and peripheral Cannabinoid1 receptors in the antihyperalgesic activity of cannabinoids in a model of neuropathic pain. Pain 2001;92:91–100.

[28] Pertwee RG. Pharmacological actions of cannabinoids. Handb Exp Pharmacol 2005;1–51.

[29] Bradshaw HB, Rimmerman N, Hu SS, Benton VM, Stuart JM, Masuda K, et al. The endocannabinoid anandamide is a precursor for the signaling lipid N-arachidonyl glycine through two distinct pathways. BMC Biochem 2009;10:14.

[30] Huang SM, Bisogno T, Petros TJ, Chang SY, Zavitsanos PA, Zipkin RE, et al. Identification of a new class of molecules, the arachidonyl amino acids, and characterization of one member that inhibits pain. J Biol Chem 2001;276:42639–44.

[31] Vuong LAQ, Mitchell VA, Vaughan CW. Actions of N-arachidonyl-glycine in a rat neuropathic pain model. Neuropharmacology 2008;54:189–93.

[32] Wiles AL, Pearlman RJ, Rosvall M, Aubrey KR, Vandenberg RJ. N-Arachidonyl-glycine inhibits the glycine transporter, GLYT2a. J Neurochem 2006;99:781–6.

[33] Yang Z, Aubrey KR, Alroy I, Harvey RJ, Vandenberg RJ, Lynch JW. Subunit-specific modulation of glycine receptors by cannabinoids and N-arachidonyl-glycine. Biochem Pharmacol 2008;76:1014–23.

[34] Bondarenko AI, Drachuk K, Panasiuk O, Sagach V, Deak AT, Malli R, et al. N-Arachidonoyl glycine suppresses Na^+/Ca^{2+} exchanger-mediated Ca^{2+} entry into endothelial cells and activates BK_{Ca} channels independently of GPCRs. Br J Pharmacol 2013;169:933–48.

[35] Bondarenko AI, Malli R, Graier WF. The GPR55 agonist lysophosphatidylinositol directly activates intermediate-conductance Ca^{2+}-activated K^+ channels. Pflugers Arch 2011;462:245–55.

[36] Kohno M, Hasegawa H, Inoue A, Muraoka M, Miyazaki T, Oka K, et al. Identification of N-arachidonylglycine as the endogenous ligand for orphan G-protein-coupled receptor GPR18. Biochem Biophys Res Commun 2006;347:827–32.

[37] McHugh D, Page J, Dunn E, Bradshaw HB. Δ^9-THC and N-arachidonyl glycine are full agonists at GPR18 and cause migration in the human endometrial cell line, HEC-1B. Br J Pharmacol 2012;165:2414–24.

[38] Baker D, Pryce G, Davies WL, Hiley CR. *In silico* patent searching reveals a new cannabinoid receptor. Trends Pharmacol Sci 2006;27:1–4.

[39] Ross RA. The enigmatic pharmacology of GPR55. Trends Pharmacol Sci 2009;30:156–63.

[40] Henstridge CM, Balenga NA, Schroder R, Kargl JK, Platzer W, Martini L, et al. GPR55 ligands promote receptor coupling to multiple signalling pathways. Br J Pharmacol 2010;160:604–14.

[41] Kapur A, Zhao P, Sharir H, Bai Y, Caron MG, Barak LS, et al. Atypical responsiveness of the orphan receptor GPR55 to cannabinoid ligands. J Biol Chem 2009;284:29817–27.

[42] Ryberg E, Larsson N, Sjogren S, Hjorth S, Hermansson NO, Leonova J, et al. The orphan receptor GPR55 is a novel cannabinoid receptor. Br J Pharmacol 2007;152:1092–101.

[43] Oka S, Nakajima K, Yamashita A, Kishimoto S, Sugiura T. Identification of GPR55 as a lysophosphatidylinositol receptor. Biochem Biophys Res Commun 2007;362:928–34.

[44] Oka S, Toshida T, Maruyama K, Nakajima K, Yamashita A, Sugiura T. 2-Arachidonoyl-*sn*-glycero-3-phosphoinositol: a possible natural ligand for GPR55. J Biochem 2009;145:13–20.

[45] Kotsikorou E, Sharir H, Shore DM, Hurst DP, Lynch DL, Madrigal KE, et al. Identification of the GPR55 antagonist binding site using a novel set of high potency GPR55 selective ligands. Biochemistry 2013;52:9456–69.

[46] Heynen-Genel S, Dahl R, Shi S, Milan L, Hariharan S, Sergienko E, et al. Screening for selective ligands for GPR55 – antagonists. Probe Reports from the NIH Molecular Libraries Program. 2010.

[47] Overton HA, Babbs AJ, Doel SM, Fyfe MC, Gardner LS, Griffin G, et al. Deorphanization of a G protein-coupled receptor for oleoylethanolamide and its use in the discovery of small-molecule hypophagic agents. Cell Metab 2006;3:167–75.

[48] Hansen KB, Rosenkilde MM, Knop FK, Wellner N, Diep TA, Rehfeld JF, et al. 2-Oleoyl glycerol is a GPR119 agonist and signals GLP-1 release in humans. J Clin Endocrinol Metab 2011;96:E1409–17.

[49] Pickel VM, Chan J, Kearn CS, Mackie K. Targeting dopamine D2 and cannabinoid-1 (CB1) receptors in rat nucleus accumbens. J Comp Neurol 2006;495:299–313.

[50] Kearn CS, Blake-Palmer K, Daniel E, Mackie K, Glass M. Concurrent stimulation of cannabinoid CB1 and dopamine D2 receptors enhances heterodimer formation: A mechanism for receptor cross-talk? Mol Pharmacol 2005;67:1697–704.

[51] Marcellino D, Carriba P, Filip M, Borgkvist A, Frankowska M, Bellido I, et al. Antagonistic cannabinoid CB1/dopamine D-2 receptor interactions in striatal CB1/D-2 heteromers. A combined neurochemical and behavioral analysis. Neuropharmacology 2008;54:815–23.

[52] Schoffelmeer AN, Hogenboom F, Wardeh G, De Vries TJ. Interactions between CB1 cannabinoid and mu opioid receptors mediating inhibition of neurotransmitter release in rat nucleus accumbens core. Neuropharmacology 2006;51:773–81.

[53] Hojo M, Sudo Y, Ando Y, Minami K, Takada M, Matsubara T, et al. mu-Opioid receptor forms a functional heterodimer with cannabinoid CB1 receptor: electrophysiological and FRET assay analysis. J Pharmacol Sci 2008;108:308–19.

[54] Bushlin I, Gupta A, Stockton SD Jr, Miller LK, Devi LA. Dimerization with cannabinoid receptors allosterically modulates delta opioid receptor activity during neuropathic pain. PLoS One 2012;7:e49789.

[55] Ellis J, Pediani JD, Canals M, Milasta S, Milligan G. Orexin-1 receptor-cannabinoid CB1 receptor heterodimerization results in both ligand-dependent and -independent coordinated alterations of receptor localization and function. J Biol Chem 2006;281:38812–24.

[56] Carriba P, Ortiz O, Patkar K, Justinova Z, Stroik J, Themann A, et al. Striatal adenosine A_{2A} and cannabinoid CB_1 receptors form functional heteromeric complexes that mediate the motor effects of cannabinoids. Neuropsychopharmacology 2007;32:2249–59.

[57] Hudson BD, Hebert TE, Kelly ME. Physical and functional interaction between CB_1 cannabinoid receptors and β_2-adrenoceptors. Br J Pharmacol 2010;160:627–42.

[58] Zygmunt PM, Petersson J, Andersson DA, Chuang HH, Sørgård M, Di Marzo V, et al. Vanilloid receptors on sensory nerves mediate the vasodilator action of anandamide. Nature 1999;400:452–7.

[59] Di Marzo V, Bisogno T, Melck D, Ross R, Brockie H, Stevenson L, et al. Interactions between synthetic vanilloids and the endogenous cannabinoid system. FEBS Lett 1998;436:449–54.

[60] De Petrocellis L, Ligresti A, Moriello AS, Allara M, Bisogno T, Petrosino S, et al. Effects of cannabinoids and cannabinoid-enriched Cannabis extracts on TRP channels and endocannabinoid metabolic enzymes. Br J Pharmacol 2011;163:1479–94.

[61] Bisogno T, Hanus L, De Petrocellis L, Tchilibon S, Ponde DE, Brandi I, et al. Molecular targets for cannabidiol and its synthetic analogues: effect on vanilloid VR1 receptors and on the cellular uptake and enzymatic hydrolysis of anandamide. Br J Pharmacol 2001;134:845–52.

[62] Qin N, Neeper MP, Liu Y, Hutchinson TL, Lubin ML, Flores CM. TRPV2 is activated by cannabidiol and mediates CGRP release in cultured rat dorsal root ganglion neurons. J Neurosci 2008;28:6231–8.

[63] De Petrocellis L, Orlando P, Moriello AS, Aviello G, Stott C, Izzo AA, et al. Cannabinoid actions at TRPV channels: effects on TRPV3 and TRPV4 and their potential relevance to gastrointestinal inflammation. Acta Physiol (Oxf) 2012;204:255–66.

[64] De Petrocellis L, Vellani V, Schiano-Moriello A, Marini P, Magherini PC, Orlando P, et al. Plant-derived cannabinoids modulate the activity of transient receptor potential channels of ankyrin type-1 and melastatin type-8. J Pharmacol Exp Ther 2008;325:1007–15.

[65] Coyne L, Lees G, Nicholson RA, Zheng J, Neufield KD. The sleep hormone oleamide modulates inhibitory ionotropic receptors in mammalian CNS *in vitro*. Br J Pharmacol 2002;135:1977–87.

[66] Sigel E, Baur R, Racz I, Marazzi J, Smart TG, Zimmer A, et al. The major central endocannabinoid directly acts at $GABA_A$ receptors. Proc Natl Acad Sci USA 2011;108:18150–5.

[67] Oz M, Ravindran A, Diaz-Ruiz O, Zhang L, Morales M. The endogenous cannabinoid anandamide inhibits α7 nicotinic acetylcholine receptor-mediated responses in *Xenopus* oocytes. J Pharmacol Exp Ther 2003;306:1003–10.

[68] Barann M, Molderings G, Bruss M, Bonisch H, Urban BW, Gothert M. Direct inhibition by cannabinoids of human 5-HT3A receptors: probable involvement of an allosteric modulatory site. Br J Pharmacol 2002;137:589–96.

[69] Fan P. Cannabinoid agonists inhibit the activation of 5-HT3 receptors in rat nodose ganglion neurons. J Neurophysiol 1995;73:907–10.

[70] Hampson AJ, Bornheim LM, Scanziani M, Yost CS, Gray AT, Hansen BM, et al. Dual effects of anandamide on NMDA receptor-mediated responses and neurotransmission. J Neurochem 1998;70:671–6.

[71] Ross HR, Napier I, Connor M. Inhibition of recombinant human T-type calcium channels by Δ^9-tetrahydrocannabinol and cannabidiol. J Biol Chem 2008;283:16124–34.

[72] Cazade M, Nuss CE, Bidaud I, Renger JJ, Uebele VN, Lory P, et al. Cross modulation and molecular interaction at the Cav3.3 protein between the endogenous lipids and the T-type calcium channel antagonist TTA-A2. Mol Pharmacol 2014;85:218–25.

[73] Barbara G, Alloui A, Nargeot J, Lory P, Eschalier A, Bourinet E, et al. T-type calcium channel inhibition underlies the analgesic effects of the endogenous lipoamino acids. J Neurosci 2009;29:13106–14.

[74] Chemin J, Monteil A, Perez-Reyes E, Nargeot J, Lory P. Direct inhibition of T-type calcium channels by the endogenous cannabinoid anandamide. EMBO J 2001;20:7033–40.

[75] O'Sullivan SE. Cannabinoids go nuclear: evidence for activation of peroxisome proliferator-activated receptors. Br J Pharmacol 2007;152:576–82.

[76] Sun Y, Alexander SPH, Kendall DA, Bennett AJ. Cannabinoids and PPARα signalling. Nottingham, UK: School of Biomedical Sciences, University of Nottingham; 2006. p. 1095–1097.

[77] Michalik L, Auwerx J, Berger JP, Chatterjee VK, Glass CK, Gonzalez FJ, et al. International Union of Pharmacology. LXI. Peroxisome proliferator-activated receptors. Pharmacol Rev 2006;58:726–41.

[78] Huang JT, Welch JS, Ricote M, Binder CJ, Willson TM, Kelly C, et al. Interleukin-4-dependent production of PPAR-γ ligands in macrophages by 12/15-lipoxygenase. Nature 1999;400:378–82.

[79] Lee G, Elwood F, McNally J, Weiszmann J, Lindstrom M, Amaral K, et al. T0070907, a selective ligand for peroxisome proliferator-activated receptor γ, functions as an antagonist of biochemical and cellular activities. J Biol Chem 2002;277:19649–57.

[80] Xu HE, Stanley TB, Montana VG, Lambert MH, Shearer BG, Cobb JE, et al. Structural basis for antagonist-mediated recruitment of nuclear co-repressors by PPARα. Nature 2002;415:813–7.

[81] Shearer BG, Steger DJ, Way JM, Stanley TB, Lobe DC, Grillot DA, et al. Identification and characterization of a selective peroxisome proliferator-activated receptor β/δ (NR1C2) antagonist. Mol Endocrinol 2008;22:523–9.

[82] Kehrer JP, Biswal SS, La E, Thuillier P, Datta K, Fischer SM, et al. Inhibition of peroxisome-proliferator-activated receptor (PPAR)α by MK886. Biochem J 2001;356:899–906.

[83] Dixon RA, Diehl RE, Opas E, Rands E, Vickers PJ, Evans JF, et al. Requirement of a 5-lipoxygenase-activating protein for leukotriene synthesis. Nature 1990;343:282–4.

[84] Panlilio LV, Justinova Z, Mascia P, Pistis M, Luchicchi A, Lecca S, et al. Novel use of a lipid-lowering fibrate medication to prevent nicotine reward and relapse: preclinical findings. Neuropsychopharmacology 2012;37:1838–47.

[85] Inestrosa NC, Carvajal FJ, Zolezzi JM, Tapia-Rojas C, Serrano F, Karmelic D, et al. Peroxisome proliferators reduce spatial memory impairment, synaptic failure, and neurodegeneration in brains of a double transgenic mice model of Alzheimer's disease. J Alzheimers Dis 2013;33:941–59.

[86] Nenov MN, Laezza F, Haidacher SJ, Zhao Y, Sadygov RG, Starkey JM, et al. Cognitive enhancing treatment with a PPARgamma agonist normalizes dentate granule cell presynaptic function in Tg2576 APP mice. J Neurosci 2014;34:1028–36.

[87] Kalonia H, Kumar P, Kumar A. Pioglitazone ameliorates behavioral, biochemical and cellular alterations in quinolinic acid induced neurotoxicity: possible role of peroxisome proliferator activated receptor-Gamma (PPAR Gamma) in Huntington's disease. Pharmacol Biochem Behav 2010;96:115–24.

[88] Jin J, Albertz J, Guo Z, Peng Q, Rudow G, Troncoso JC, et al. Neuroprotective effects of PPAR-gamma agonist rosiglitazone in N171-82Q mouse model of Huntington's disease. J Neurochem 2013;125:410–9.

[89] Carta AR, Frau L, Pisanu A, Wardas J, Spiga S, Carboni E. Rosiglitazone decreases peroxisome proliferator receptor-gamma levels in microglia and inhibits Tnf-alpha production: new evidences on neuroprotection in a progressive Parkinson's disease model. Neuroscience 2011;194:250–61.

[90] Puligheddu M, Pillolla G, Melis M, Lecca S, Marrosu F, De Montis MG, et al. PPAR-alpha agonists as novel antiepileptic drugs: preclinical findings. PLoS One 2013;8:e64541.

[91] Hong S, Xin Y, Haiqin W, Guilian Z, Ru Z, Shuqin Z, et al. The PPARgamma agonist rosiglitazone prevents cognitive impairment by inhibiting astrocyte activation and oxidative stress following pilocarpine-induced status epilepticus. Neurol Sci 2012;33:559–66.

[92] Gray E, Ginty M, Kemp K, Scolding N, Wilkins A. The PPAR-gamma agonist pioglitazone protects cortical neurons from inflammatory mediators via improvement in peroxisomal function. J Neuroinflammation 2012;9:63.

[93] Bouaboula M, Hilairet S, Marchand J, Fajas L, Le Fur G, Casellas P. Anandamide induced PPARγ transcriptional activation and 3T3-L1 preadipocyte differentiation. Eur J Pharmacol 2005;517:174–81.

[94] Fu J, Gaetani S, Oveisi F, Lo Verme J, Serrano A, Rodriguez De Fonseca F, et al. Oleylethanolamide regulates feeding and body weight through activation of the nuclear receptor PPAR-α. Nature 2003;425:90–3.

[95] O'Sullivan SE, Tarling EJ, Bennett AJ, Kendall DA, Randall MD. Novel time-dependent vascular actions of Δ^9-tetrahydrocannabinol mediated by peroxisome proliferator-activated receptor γ. Biochem Biophys Res Commun 2005;337:824–31.

[96] O'Sullivan SE, Sun Y, Bennett AJ, Randall MD, Kendall DA. Time-dependent vascular actions of cannabidiol in the rat aorta. Eur J Pharmacol 2009;612:61–8.

[97] Guzmán M, Lo Verme J, Fu J, Oveisi F, Blázquez C, Piomelli D. Oleoylethanolamide stimulates lipolysis by activating the nuclear receptor peroxisome proliferator-activated receptor α (PPAR-α). J Biol Chem 2004;279:27849–54.

[98] Lo Verme J, Fu J, Astarita G, La Rana G, Russo R, Calignano A, et al. The nuclear receptor peroxisome proliferator-activated receptor-α mediates the antiinflammatory actions of palmitoylethanolamide. Mol Pharmacol 2005;67:15–9.

[99] Carroll CB, Zeissler ML, Hanemann CO, Zajicek JP. Δ^9-Tetrahydrocannabinol (Δ^9-THC) exerts a direct neuroprotective effect in a human cell culture model of Parkinson's disease. Neuropathol Appl Neurobiol 2012;38:535–47.

[100] Ramer R, Heinemann K, Merkord J, Rohde H, Salamon A, Linnebacher M, et al. COX-2 and PPAR-γ confer cannabidiol-induced apoptosis of human lung cancer cells. Mol Cancer Ther 2013;12:69–82.

[101] O'Sullivan SE, Kendall DA, Randall MD. Further characterization of the time-dependent vascular effects of Δ^9-tetrahydrocannabinol. J Pharmacol Exp Ther 2006;317:428–38.

[102] O'Sullivan SE, Kendall DA, Randall MD. Time-dependent vascular effects of endocannabinoids mediated by Peroxisome Proliferator-Activated Receptor Gamma (PPARγ). PPAR Res 2009;2009:425289.

[103] Scuderi C, Valenza M, Stecca C, Esposito G, Carratu MR, Steardo L. Palmitoylethanolamide exerts neuroprotective effects in mixed neuroglial cultures and organotypic hippocampal slices via peroxisome proliferator-activated receptor-α. J Neuroinflammation 2012;9:49.

[104] Kumar A, Qiao Z, Kumar P, Song ZH. Effects of palmitoylethanolamide on aqueous humor outflow. Invest Ophthalmol Vis Sci 2012;53:4416–25.

[105] Costa B, Comelli F, Bettoni I, Colleoni M, Giagnoni G. The endogenous fatty acid amide, palmitoylethanolamide, has anti-allodynic and anti-hyperalgesic effects in a murine model of neuropathic pain: involvement of CB_1, TRPV1 and PPARγ receptors and neurotrophic factors. Pain 2008;139:541–50.

[106] de Novellis V, Luongo L, Guida F, Cristino L, Palazzo E, Russo R, et al. Effects of intra-ventrolateral periaqueductal grey palmitoylethanolamide on thermoceptive threshold and rostral ventromedial medulla cell activity. Eur J Pharmacol 2012;676:41–50.

[107] Fakhfouri G, Ahmadiani A, Rahimian R, Grolla AA, Moradi F, Haeri A. WIN55212-2 attenuates amyloid-beta-induced neuroinflammation in rats through activation of cannabinoid receptors and PPAR-γ pathway. Neuropharmacology 2012;63:653–66.

[108] Luchicchi A, Lecca S, Carta S, Pillolla G, Muntoni AL, Yasar S, et al. Effects of fatty acid amide hydrolase inhibition on neuronal responses to nicotine, cocaine and morphine in the nucleus accumbens shell and ventral tegmental area: involvement of PPAR-α nuclear receptors. Addict Biol 2010;15:277–88.

[109] Kaczocha M, Vivieca S, Sun J, Glaser ST, Deutsch DG. Fatty acid binding proteins transport N-acylethanolamines to nuclear receptors and are targets of endocannabinoid transport inhibitors. J Biol Chem 2012;287:3415–24.

[110] Newberry EP, Kennedy SM, Xie Y, Luo J, Crooke RM, Graham MJ, et al. Decreased body weight and hepatic steatosis with altered fatty acid ethanolamide metabolism in aged L-Fabp$^{-/-}$ mice. J Lipid Res 2012;53:744–54.

[111] Sun Y, Alexander S, Kendall D, Bennett A. FABPs deliver FAEs signalling to PPARs. 2007. p. PC284.

[112] Lee SS, Pineau T, Drago J, Lee EJ, Owens JW, Kroetz DL, et al. Targeted disruption of the alpha isoform of the peroxisome proliferator-activated receptor gene in mice results in abolishment of the pleiotropic effects of peroxisome proliferators. Mol Cell Biol 1995;15:3012–22.

[113] Peters JM, Lee SS, Li W, Ward JM, Gavrilova O, Everett C, et al. Growth, adipose, brain, and skin alterations resulting from targeted disruption of the mouse peroxisome proliferator-activated receptor β(δ). Mol Cell Biol 2000;20:5119–28.

[114] Barak Y, Liao D, He W, Ong ES, Nelson MC, Olefsky JM, et al. Effects of peroxisome proliferator-activated receptor delta on placentation, adiposity, and colorectal cancer. Proc Natl Acad Sci USA 2002;99:303–8.

[115] Barak Y, Nelson MC, Ong ES, Jones YZ, Ruiz-Lozano P, Chien KR, et al. PPAR gamma is required for placental, cardiac, and adipose tissue development. Mol Cell 1999;4:585–95.

[116] Rosen ED, Sarraf P, Troy AE, Bradwin G, Moore K, Milstone DS, et al. PPARγ is required for the differentiation of adipose tissue *in vivo* and *in vitro*. Mol Cell 1999;4:611–7.

[117] Kubota N, Terauchi Y, Miki H, Tamemoto H, Yamauchi T, Komeda K, et al. PPAR gamma mediates high-fat diet-induced adipocyte hypertrophy and insulin resistance. Mol Cell 1999;4:597–609.

[118] Akiyama TE, Sakai S, Lambert G, Nicol CJ, Matsusue K, Pimprale S, et al. Conditional disruption of the peroxisome proliferator-activated receptor γ gene in mice results in lowered expression of ABCA1, ABCG1, and apoE in macrophages and reduced cholesterol efflux. Mol Cell Biol 2002;22:2607–19.

[119] LoVerme J, Russo R, La Rana G, Fu J, Farthing J, Mattace-Raso G, et al. Rapid broad-spectrum analgesia through activation of peroxisome proliferator-activated receptor-α. J Pharmacol Exp Ther 2006;319:1051–61.

[120] D'Agostino G, La Rana G, Russo R, Sasso O, Iacono A, Esposito E, et al. Acute intracerebroventricular administration of palmitoylethanolamide, an endogenous PPAR-α agonist, modulates carrageenan-induced paw edema in mice. J Pharmacol Exp Ther 2007;322:1137–43.

[121] D'Agostino G, La Rana G, Russo R, Sasso O, Iacono A, Esposito E, et al. Central administration of palmitoylethanolamide reduces hyperalgesia in mice via inhibition of NF-kB nuclear signalling in dorsal root ganglia. Eur J Pharmacol 2009;613:54–9.

[122] Sun Y, Alexander SPH, Garle MJ, Gibson CL, Hewitt K, Murphy SP, et al. Cannabinoid activation of PPARα: a novel neuroprotective mechanism. Br J Pharmacol 2007;152:734–43.

[123] Suardíaz M, Estivill-Torrús G, Goicoechea C, Bilbao A, Rodríguez de Fonseca F. Analgesic properties of oleoylethanolamide (OEA) in visceral and inflammatory pain. Pain 2007;133:99–110.

[124] Cluny NL, Keenan CM, Lutz B, Piomelli D, Sharkey KA. The identification of peroxisome proliferator-activated receptor alpha-independent effects of oleoylethanolamide on intestinal transit in mice. Neurogastroenterology and Motility 2009;21:420–9.

[125] Melis M, Pillolla G, Luchicchi A, Muntoni AL, Yasar S, Goldberg SR, et al. Endogenous fatty acid ethanolamides suppress nicotine-induced activation of mesolimbic dopamine neurons through nuclear receptors. J Neurosci 2008;28:13985–94.

[126] Jhaveri MD, Richardson D, Robinson I, Garle MJ, Patel A, Sun Y, et al. Inhibition of fatty acid amide hydrolase and cycloxygenase-2 increases levels of endocannabinoids and produces analgesia via peroxisome proliferator-activated receptor-alpha in a model of inflammatory pain. Neuropharmacology 2008;55:85–93.

[127] Sagar DR, Kendall DA, Chapman V. Inhibition of fatty acid amide hydrolase produces PPAR-α-mediated analgesia in a rat model of inflammatory pain. Br J Pharmacol 2008;155:1297–306.

[128] Loría F, Petrosino S, Hernangómez M, Mestre L, Spagnolo A, Correa F, et al. An endocannabinoid tone limits excitotoxity *in vitro* and in a model of multiple sclerosis. Neurobiol Dis 2010;37:166–76.

[129] Roche M, Kelly JP, O'Driscoll M, Finn DP. Augmentation of endogenous cannabinoid tone modulates lipopolysaccharide-induced alterations in circulating cytokine levels in rats. Immunology 2008;125:263–71.

[130] Khasabova IA, Xiong Y, Coicou LG, Piomelli D, Seybold V. Peroxisome proliferator-activated receptor α mediates acute effects of palmitoylethanolamide on sensory neurons. J Neurosci 2012;32:12735–43.

[131] Sasso O, Moreno-Sanz G, Martucci C, Realini N, Dionisi M, Mengatto L, et al. Antinociceptive effects of the *N*-acylethanolamine acid amidase inhibitor ARN077 in rodent pain models. Pain 2013;154:350–60.

[132] Kozak KR, Gupta RA, Moody JS, Ji C, Boeglin WE, DuBois RN, et al. 15-Lipoxygenase metabolism of 2-arachidonylglycerol: generation of a peroxisome proliferator-activated receptor α agonist. J Biol Chem 2002;277:23278–86.

[133] Fang X, Hu SM, Xu BK, Snyder GD, Harmon S, Yao JR, et al. 14,15-Dihydroxyeicosatrienoic acid activates peroxisome proliferator-activated receptor-α. Am J Physiol – Heart Circ Physiol 2006;290:H55–63.

[134] Ghosh M, Wang H, Ai Y, Romeo E, Luyendyk JP, Peters JM, et al. COX-2 suppresses tissue factor expression via endocannabinoid-directed PPARδ activation. J Exp Med 2007;204:2053–61.

[135] Forman BM, Chen J, Evans RM. Hypolipidemic drugs, polyunsaturated fatty acids, and eicosanoids are ligands for peroxisome proliferator-activated receptors α and δ. Proc Natl Acad Sci USA 1997;94:4312–7.

Index

Printed in the ... ...
B. ... ...

Printed in the United States
By Bookmasters